T0311582

DESIGNING ENTERPRISE ARCHITECTURE FRAMEWORKS

Integrating Business Processes
with IT Infrastructure

DESIGNING ENTERPRISE ARCHITECTURE FRAMEWORKS

Integrating Business Processes with IT Infrastructure

Edited by
Liviu Gabriel Cretu, PhD

Apple Academic Press

TORONTO   NEW JERSEY

Apple Academic Press Inc. | Apple Academic Press Inc.
3333 Mistwell Crescent | 9 Spinnaker Way
Oakville, ON L6L 0A2 | Waretown, NJ 08758
Canada | USA

©2014 by Apple Academic Press, Inc.

First issued in paperback 2021

Exclusive worldwide distribution by CRC Press, a member of Taylor & Francis Group
No claim to original U.S. Government works

ISBN 13: 978-1-77463-329-8 (pbk)
ISBN 13: 978-1-77188-007-7 (hbk)

Library of Congress Control Number: 2013954453

Library and Archives Canada Cataloguing in Publication

Designing enterprise architecture frameworks: integrating business processes with IT infrastructure/edited by Liviu Gabriel Cretu, PhD.

Includes bibliographical references and index.
ISBN 978-1-77188-007-7
1. Management information systems. 2. Business--Data processing.
3. Information technology--Management. 4. Computer network architectures.
I. Cretu, Liviu Gabriel, editor of compilation

HD30.213.D48 2014   658.4'038011   C2013-907653-0

ABOUT THE EDITOR

Liviu Gabriel Cretu, PhD

Dr. Liviu Gabriel Cretu is a research director at the Centre for Research and Education in Management Information Systems, Alexandru Ioan Cuza University, Iaşi, Romania; a lecturer at the same university; and a consultant in applications architecture at the European Commission, D.G. Informatics, Unit B-Information Systems and Interoperability Solutions. He is a researcher, developer, and author, having published numerous peer-reviewed articles in the fields of enterprise systems, semantic web and enterprise architectures, model-driven engineering, service-oriented architecture, and business process management. He received his PhD in business information systems from the Alexandru Ioan Cuza University, Iaşi, Romania.

CONTENTS

ACKNOWLEDGMENT AND
HOW TO CITE

The chapters in this book were previously published in various places and in various formats. By bringing them together here in one place, we offer the reader a comprehensive perspective on recent investigations into the design and implementation of enterprise architecture frameworks. Each chapter is added to and enriched by being placed within the context of the larger investigative landscape. Specifically:

- Chapter 1 provides one of the few, well-documented overviews of the main enterprise architecture frameworks. The authors have also made a distinct contribution to the field by proposing a comparison method based on higher order goals as well as non-functional requirements. It is a very good starting point in order to understand the EA landscape.
- Chapter 2 makes a very well-documented and in-depth comparison between two highly popular EA frameworks: ARIS and Zachman. The level of detail as well as the accuracy of the analysis are rather high, thus transforming the paper into a "must-read" for the beginner to better understand the domain of enterprise architectures.
- Chapter 3 proposes an interesting approach to compare well-known enterprise architecture frameworks in order to build an integrated EA model needed to represent interorganizational concerns. The main goal is to identify the common elements needed to facilitate the communication between companies using different frameworks to represent their enterprise architecture.
- Why is enterprise architecture important for business and which precise business areas might EA help improve? The answer to this question is not only given but also very well documented in Chapter 4.
- Chapter 5 introduces a novel idea of embedding patterns in enterprise architecture frameworks, in order to create new architectures based on pre-defined business and IT building blocks. The authors identify ten architecture design patterns, one for each of the TOGAF's main activities.
- Chapter 6 details a new content framework and metamodel for enterprise architecture with the aim to reduce the number of architecture layers and to organize the content of each architecture layer. The main contribution of this paper is the artifacts needed to describe EAs. The authors also compare their approach with well-known EAFs.

- Chapter 7 offers a different perspective on enterprise architecture than the classical, heavy-weighted EA frameworks. The authors propose a combination of EA with agent-oriented architecture, which should address the ever-changing requirements in managing complex information systems.
- Chapter 8 proposes a different classification of the EA views, namely people, processes, technology and data. The authors compare their framework with the Zachman framework and show how the ARIS toolset may be used to create an ADaPPT enterprise architecture.
- Although the method provided in Chapter 9 might be improved, still the subject of EA evaluation has an intrinsic value of its own. Architecture evaluation is key aspect when deciding which approach to take in order to move a system from "as-is" state to the desired new state.
- Chapter 10 presents a rigorous approach for the transition between "as-is" enterprise model to the "to be" enterprise architecture. The authors combine the famous Motorola's Six Sigma method for process improvement with the enterprise architecture framework, having the main goal to guide the transformation process of the enterprise model.
- Chapter 11 provides a study on the status of architecture implementations within the industry, also taking into account the level of implementation maturity as well as statistical analysis of the frameworks used by various companies with respect to their operating model.
- Chapter 12 details in-depth study of EA adoption issues with relation to software development methodologies. It focuses on the need of some architecture methodology in any company, and it mainly analyzes the SEAM framework implementation.
- Chapter 13 uses an ontology-based approach to align maintenance and repair supply chain management to the ISO 10303 standard. Since this process involves multiple organizations, the common understanding of the whole model is built using the Zachman framework and correlation ontology.
- Chapter 14 uses an EA framework to organize the core components specifically needed in mobile-businesses. The work provides an EA usage scenario for the specific domain model.
- Chapter 15 is an interesting and pragmatic usage of EA in combination with model-driven architecture (MDA) to solve particular issues in the conception and management of information systems for the higher education sector. It also offers a case study of the transition between "as-is" to the "to-be" architecture of a specific university.

We wish to thank the authors who made their research available for this book, whether by granting permission individually or by releasing their research as open source articles. When citing information contained within this book, please do the authors the courtesy of attributing them by name, referring back to their original articles, using the credits provided at the beginning of each chapter.

LIST OF CONTRIBUTORS

Kamran Ahsan
Faculty of Computing, Engineering and Technology, Staffordshire University, Stafford, United Kingdom

Keijiro Araki
Dept. of Advanced Information Technology, Kyushu University, Fukuoka, Japan

Bouchaib Bounabat
Al Qalasadi Computer Science Laboratory, ENSIAS, Université Mohammed V – Souissi BP 713, Agdal – Rabat, Morocco

Sekhar Chattopadhyay
RMIT University; Australia

Omar Cherkaoui
University of Quebec at Montreal, Montreal, Canada

Zulkhairi Md. Dahalin
Utara Malaysia, Sintok, Malaysia

M. de Vries
Department of Industrial and Systems Engineering University of Pretoria, South Africa

Shantanu K. Dixit
Department of Electronics and Telecommunications, WIT, Solapur, India

Mahesh R. Dube
Department of Computer Engineering, VIT, Pune, India

Paul Golder
Birmingham City University, TEE, Millennium Point, Birmingham, B4 7XG, UK

Khawar Hameed
Faculty of Computing, Engineering and Technology, Staffordshire University, Stafford, United Kingdom

Huda Ibrahim
Utara Malaysia, Sintok, Malaysia

M. Khairudin Kasiran
Utara Malaysia, Sintok, Malaysia

Samaneh Khamseh
Department of Computer Engineering, Science and Research Branch, Islamic Azad University, Ilam, Iran

Melita Kozina
University of Zagreb, Faculty of Organization and Informatics, Varaždin, Croatia

Mouhsine Lakhdissi
Al Qalasadi Computer Science Laboratory, ENSIAS, Université Mohammed V – Souissi BP 713, Agdal – Rabat, Morocco

Mardiana
Dept. of Advanced Information Technology, Kyushu University, Fukuoka, Japan, and Dept. of Electrical Engineering, University of Lampung, Indonesia

John P. T. Mo
RMIT University, Australia

Fatemeh Nikpay
Advanced Informatics School, University Technology Malaysia (UTM),
Kuala Lumpur, Malaysia

Pankaj Pankaj
MIS and Decision Sciences, Eberly College of Business & Information Technology, Indiana University of Pennsylvania, Indiana County, Pennsylvania, USA

Rafidah Abd Razak
Utara Malaysia, Sintok, Malaysia

A. C. J van Rensburg
Department of Industrial and Systems Engineering University of Pretoria, South Africa

James A. Rodger
MIS and Decision Sciences, Eberly College of Business & Information Technology, Indiana University of Pennsylvania, Indiana County, Pennsylvania, USA

Babak Darvish Rouhani
Advanced Informatics School, University Technology Malaysia (UTM) Kuala Lumpur, Malaysia

Fares Sayyadi
Faculty Member of Electrical and Science, Science and Research Branch, Islamic Azad University, Ilam, Iran

Hanifa Shah
Birmingham City University, TEE, Millennium Point, Birmingham, B4 7XG, UK

Mohamed Taleb
École de Technologie Supérieure (ÉTS), Montreal, Canada

Raymond Vella
Ford Motor Company, Australia

Roel Wieringa
University of Twente, Department of Computer Science, Information Systems Group P.O. Box 217, 7500 AE Enschede, The Netherlands

Weijun Yang
Faculty of Computing, Engineering and Technology, Staffordshire University, Stafford, United Kingdom

Mohammad Hosein Yektaie
Faculty Member of Computer and Science, Abadan branch, Islamic Azad University, Abadan, Iran

Nor Iadah Yusop
Utara Malaysia, Sintok, Malaysia

Novica Zarvić
University of Twente, Department of Computer Science, Information Systems Group P.O. Box 217, 7500 AE Enschede, The Netherlands

INTRODUCTION

Large software systems are defined as systems where the size is measured in millions or tens of millions of lines of code and the life of the project is measured in years [1]. Complex software is usually associated with enterprise information systems and it is built to support all the business processes of an organization. Large software systems involve inherent complexity, which does not make their development and maintenance an easy task. Brown [2] names the following difficulties arising in the development of enterprise software solutions: (1) understanding highly complex business domains and management of the huge development effort; (2) time-to-market pressures; (3) complexity of target software platforms frequently kept with poor documentation. To address these issues, modern software engineering employs tools and techniques specifically designed to manage the complexity of systems. Among these, enterprise architecture (EA) frameworks are the most advanced management tools to organize various models describing an organization and to align the business needs to software components in such a way that all the stakeholders have the same understanding of the whole enterprise system.

In the 1987 article "A Framework for Information Systems Architecture" [3] John Zachman proposed a classification schema for organizing the architecture of information systems. He started from the observation that the term "architecture" was used loosely by information systems professionals and meant different things to planners, designers, programmers, communication specialists, and others. Therefore he designed the first enterprise architecture framework that mainly provides a matrix with a synoptic view over the models needed for enterprise architecture. In this matrix one row represents a complete functional (sub-) system only regarded from the specific stakeholder's perspective. Each column represents a complete part of the system. The idea is that a translation mechanism should exist to transform the models from one row to the other.

Twenty years later the proposed EA schema proved to be still valid for the organization of IT resources in modern companies [4]. In 2007 the ISO standardization body adopted ISO/IEC 42010:2007 standard (Recommended Practice for Architectural Description of Software-intensive Systems) formerly an IEEE standard [5], which gives a formal definition to architecture descriptions: a document, repository or collection of artifacts used to define and document architectures. According to this standard, every system is considered in the context of its environment. The environment of a system is understood through the identification of the stakeholders (e.g. client for the system, users, operators, developers, suppliers, regulators) of the system and their system concerns (e.g. data structure, behavior, data access, control, cost, safety, security). In order to take into consideration both the stakeholders and the many concerns of a system, the standard introduces two fundamental constructs of the system's architecture: viewpoints and views. These can be shortly described as follows:

- A viewpoint captures the conventions for constructing, interpreting and analyzing a particular kind of view such as languages, notations, model types, modeling methods, analysis techniques, design rules and any associated methods. Examples of viewpoints include: business, conceptual, technical, physical;
- A view is a collection of models representing the architecture of the whole system relative to a set of architectural concerns. A view is part of a particular architecture description for a system of interest. Examples of views: data, functions, events, roles.

Many formal models of enterprise architecture have been developed over time and some of them became very popular in the IT industry (e.g. TOGAF, ARIS). All of them have in common the pursuit to create, manage, and evolve repositories of models, the well-designed plans of large organizations containing multiple descriptions to show different views and viewpoints of the same enterprise.

This book aims to gather the critical body of knowledge produced in the scientific literature regarding what EA is and how it may be used to better organize the descriptions of enterprise systems and to achieve the desired level of business-IT alignment. The papers have been selected in such way as to provide a solid foundation for a cross-disciplinary professional practice and to serve IT professionals with easy-to-read and under-standable materials. The book is further organized in three main sections: 1) Setting the Stage; 2) Open Issues and Novel Ideas; 3) Implementations. For each article a short introduction is given to help the reader understand the rationale behind the selection and the place of that chapter within the whole picture. To see these explanations for each chapter, please view the acknowledgment page in the front of the book.

The first section comprises four papers which together make a good introduction to the field in order to create the right level of critical knowl-edge required to understand the EA concept and its place within the land-scape of information systems management and the business-IT alignment. The reader will also become familiar with the popular frameworks that may be an option when choosing how to implement a specific EA.

Chapter 1, by Dube and Dixit, provides an introduction to enterprise architectures. Enterprise architecture defines the overall form and function of systems across an enterprise involving the stakeholders and providing a framework, standards, and guidelines for project-specific architectures. Project-specific architecture defines the form and function of the systems in a project or program within the context of the enterprise as a whole with broad scope and business alignments. Application-specific architecture defines the form and function of the applications that will be developed to realize functionality of the system with narrow scope and technical align-ments. Because of the magnitude and complexity of any enterprise inte-gration project, a major engineering and operations planning effort must be accomplished prior to any actual integration work. As the needs and the requirements vary depending on their volume, the entire enterprise problem can be broken into chunks of manageable pieces. These pieces can be implemented and tested individually with high integration effort. Therefore it becomes essential to analyze the economic and technical fea-sibility of realizable enterprise solution. It is difficult to migrate from one technological and business aspect to other as the enterprise evolves. The existing process models in system engineering emphasize on life-cycle management and low-level activity coordination with milestone verifica-tion. Many organizations are developing enterprise architecture to provide a clear vision of how systems will support and enable their business. The paper proposes an approach for selection of suitable enterprise architecture depending on the measurement framework. The framework consists of unique combination of higher order goals, non-functional requirement support and inputs-outcomes pair evaluation. The earlier efforts in this regard were concerned about only custom scales indicating the availability of a parameter in a range.

Chapter 2, by Kozina details the significance of integral business systems based on closer alignment of information technology to business processes. Comprehensive business frameworks are necessary to capture the entire complexity of such systems. These frameworks, called enterprise architectures, can provide the conceptual foundation necessary for building and managing the integral business system and all its components. The goal of this paper was to analyze the Architecture of Integrated Information Systems (ARIS) and the Zachman frameworks, to define the criteria for comparison and evaluation of these approaches, and determine their level of complement. Furthermore, the contents of the paper define the generic model of business system management supported by said concepts (frameworks) and analyzes their orientation towards value.

Zarvic and Wieringa describe in Chapter 3 how when different businesses want to integrate part of their processes and IT they need to relate their enterprise architecture frameworks. An enterprise architecture framework (EAF) is a conceptual framework for describing the architecture of a business and its information technology (IT), and their alignment. In this paper, the authors provide an integration among some well-known EAFs (Zachman, Four-domain, TOGAF and RM-ODP) and produce an integrated EAF (IEAF) that can be used as common framework to communicate about EAFs of different businesses and relate them to each other.

In Chapter 4, by de Vries and van Rensburg, the authors detail how organizations today are characterized by conglomerate organization structures that evolve through mergers and acquisitions. Corporate offices need to add superior knowledge and skills to ensure that the collection of diverse businesses is operating as more than independent units. A new management approach is required to create synergies between the diverse businesses, their processes, and system landscapes. Enterprise architecture (EA) creates value on a corporate level by facilitating process/information technology alignment and synergy between different strategic business units (SBUs). Unfortunately many EA implementations seemed to fail owing to a short-term financial focus and measurement. This article explores the possibilities of linking EA to a corporate balanced scorecard (BSC) to demonstrate its long-term financial improvement capabilities in supporting the business strategy. The aim is to use the corporate BSC context to direct EA objectives in creating contextualized value for a specific enterprise.

The next six articles arranged in section two introduce novel ideas regarding the organization, transformation, and evaluation of EA models. The integration of multiple business models using EAs as well as the content of these models are also among the subjects treated by the authors. This section seeks to draw attention on open issues in the EA field regardless of the EA framework taken into consideration for a specific purpose.

Taleb and Cherkaoui suggest in Chapter 5 that developers must be able to reuse proven solutions emerging from the best design practices to solve common design problems while composing patterns to create reusable designs that can be mapped to different types of enterprise frameworks and architectures such as The Open Group Architecture Framework (TOGAF). Without this, business analysts, designers, and developers are not properly applying design solutions or take full benefit of the power of patterns as reuse blocks, resulting in poor performance, poor scalability, and poor usability. Furthermore, these professionals may "reinvent the wheel" when attempting to implement the same design for different types of architectures of TOGAF framework. In this paper, the authors introduce different categories of design patterns as a vehicle for capturing and reusing good analyses, designs and implementation applied to TOGAF framework while detailing a motivating exemplar on how design patterns can be composed to create generic types of architectures of TOGAF framework. Then, they discuss why patterns are a suitable for developing and documenting various architectures including enterprise architectures as TOGAF.

Chapter 6, by Lakhdiss and Bounabat, argue that IS strategic planning and enterprise architecture are two major disciplines in IT architecture and governance. They pursue the same objectives and have much in common. While ISSP has benefited from business strategic planning methods and techniques, it has not evolved much since the 90s and lacks from formal, tooled and standard methodology. In the other hand, enterprise architecture has known a very fast progression in the last years, helped by market's needs and research in the domain of entreprise modeling. The basic component underlying both fields is the content framework and metamodel necessary to describe existing and future states. The aim of this paper is to present a new EA content framework and metamodel taking into consideration ISSP concerns and bridges the gap between these two fields

In Chapter 7, Rouhani and Nikpay argue that nowadays, utilizing EA by enterprises with medium-and long-term goals causes improvement in their productivity and competitiveness. With respect to varied changes in enterprise's business activities and attitudes, flexibility in information systems of EA is a crucial factor. An Agent's capacities in implementation of complex systems goal convinces huge enterprises to use agent-oriented architecture in their EA programs. Combination of EA and agent-oriented architecture introduces a new attitude in order to make better conditions for huge enterprises with complex information systems. This paper firstly enumerates the current problems of enterprise architecture, and then agent-oriented enterprise architecture is introduced as a comprehensive solution for eliminating mentioned defects deals raised. The main results of agent-oriented enterprise architecture includes: more flexibility in organizational change, reengineering organizational processes, and comprehensive coverage of all activities of huge and complex organizations with no other lateral requirements.

Enterprises have architecture; whether it is visible or invisible is another matter. Shah and Golder show in Chapter 9 how an enterprises' architecture determines the way in which it works to deliver its business objectives and the way in which it can change to continue to meet its evolving business objectives. Enterprise architectural thinking can facilitate effective strategic planning and information systems development. This paper reviews enterprise architecture (EA) and its concepts. It briefly considers EA frameworks. It describes the ADaPPT (Aligning Data, People, Processes and Technology) EA approach as a means to managing organizational complexity and change. Future research directions are discussed.

These days, we see many organizations with extremely complex systems with various processes, organizational units, individuals, and information technology support where there are complex relationships among their various elements. In these organizations, poor architecture reduces efficiency and flexibility. Enterprise architecture, with full description of the functions of information technology in the organization, attempts to reduce the complexity of the most efficient tools to reach organizational objectives. Enterprise architecture can better assess the optimal conditions for achieving organizational goals. For evaluating enterprise architecture, executable models need to be applied. Executable models using a static architectural view to describe necessary documents need to be created. Therefore, to make an executable model, we need a requirement to produce products of the enterprise architecture to create an executable model. In Chapter 9, Khamsey and colleagues show that for the production of an enterprise architecture, the object-oriented approach is implemented. The authors present an algorithm to use stereotypes by considering reliability assessment. The approach taken in this algorithm is to improve the reliability by considering additional components in parallel and using redundancy techniques to maintain the minimum number of components. Furthermore, they implement the proposed algorithm on a case study and the results are compared with previous algorithms.

In Chapter 10, Vella and colleagues show that enterprise architecture methods provide a structured system to understand enterprise activities. However, existing enterprise modeling methodologies take static views of the enterprise and do not naturally lead to a path of improvement during enterprise model transformation. This paper discusses the need for a methodology to facilitate changes for improvement in an enterprise. The six sigma methodology is proposed as the tool to facilitate progressive and continual Enterprise Model Transformation to allow businesses to adapt to meet increased customer expectation and global competition. An alignment of six sigma with phases of GERAM life cycle is described with inclusion of Critical-To-Satisfaction (CTS) requirements. The synergies of combining the two methodologies are presented in an effort to provide a more culturally embedded framework for enterprise model transformation that builds on the success of six sigma.

The rest of the material gathers various case studies and implementation scenarios presented from the perspective of different stakeholders. The aim is to illustrate the adoption of EA and its usage at the industry level.

Chapter 11, by de Vries and van Rensburg, details enterprise architecture (EA): a new discipline that has emerged from the need to create a holistic view of an enterprise, and thereby to discover business/IT integration and alignment opportunities across enterprise structures. Previous EA value propositions that merely focus on IT cost reductions will no longer convince management to invest in EA. Today, EA should enable business strategy in the organization to create value. This resides in the ability to do enterprise optimization through process standardization and integration. In order to do this, a new approach is required to integrate EA into the strategy planning process of the organization. This article explores the use of three key artifacts—operating models, core diagrams, and an operating maturity assessment as defined by Ross, Weill and Robertson [1]—as the basis of this new approach. Action research is applied to a research group to obtain qualitative feedback on the practicality of the artifacts.

Chapter 12, by Dahalin and colleagues, proposes the use of an enterprise architecture methodology known as the Systemic Enterprise Architecture Methodology (SEAM) to determine the relevance of EA in addressing the business-IT alignment. A construct that characterized EA was developed based on review of the literature. A theoretical framework build upon the SEAM was used based on a business-IT alignment market, in which supplier business systems compete to provide a value to an adopter business system. Data was empirically gathered based on survey respondents who are concerned with the adoption, planning, and implementation of EA in their organizations. Respondents were managers and executives representing the IT and senior level management of public and private organizations in Malaysia. The data collected was then analyzed based on the following factors: (1) EA business issues; (2) EA environment; (3) EA governance; and (4) EA methods, tools and frameworks. Comparative analysis was carried out based on the four factors to examine the trend and status of EA adoption and implementation in Malaysia vis-à-vis the international scenario. Statistical analysis was used to validate the SEAM, which was found to be relevant in addressing the business-IT alignment.

Ontologies have emerged as an important tool in the enterprise architecture discipline to provide the theoretical foundations for designing and representing the enterprise as a whole or a specific area or domain, in a scientific fashion. In Chapter 13, Rodger and Pankaj examine the domain of maintenance, repair, and overhaul (MRO) of the Sikorsky UH-60 helicopter involving multiple enterprises, and represents it through an ontology using the OWL Language and Protégé tool. The resulting ontology gives a formal and unambiguous model/representation of the MRO domain that can be used by multiple parties to arrive at a common shared conceptualization of the MRO domain. The ontology is designed to be conformant to ISO 13030 or the Product Life Cycle Support Standard (PLCS) standard, hence representing the state of being as per this standard especially at the interfaces between enterprises while incorporating existing reality to the greatest possible extent within the enterprises. As a result the ontology can be used to design Information Systems (IS) and their interfaces in all enterprises engaged in MRO to alleviate some of the issues present in the MRO area and to support business intelligence efforts.

The increasing deployment of mobile technologies across industry sectors is creating fertile ground for organizations to exploit new revenue streams generated from applications that exploit the mobile ecosystem. M-Commerce (mobile commerce) has been recognized as a key driving force of next generation computing, and industry analysts such as IDC have predicted revenue growth arising from m-commerce to far exceed US$27 billion by the end of the decade [1]. Mobility has, without doubt, underpinned the current wave and generation of computing systems resulting in the concept and practicality of mobile solutions becoming embedded as natural or inherent ones that support the daily functions of individuals and corporations. Chapter 14, by Hameed and colleagues, explores an approach to encapsulating the m-commerce ecosystem through the perspective of an enterprise architecture framework.

In Chapter 15, Mardiana and Araki illustrate that higher education institutions require a proper standard and model that can be implemented to enhance alignment between business strategy and existing information technologies. Developing the required model is a complex task. A combination of the EA, MDA, and SOA concepts can be one of the solutions to overcome the complexity of building a specific information technology architecture for higher education institutions. EA allows for a comprehensive understanding of the institution's main business process while defining the information system that will assist in optimizing the business process. EA essentially focuses on strategy and integration. MDA relies on models as its main element and places focuses on efficiency and quality. SOA, on the other hand, uses services as its principal element and focuses on flexibility and reuse. This paper seeks to formulate an information technology architecture that can provide clear guidelines on inputs and outputs for EA development activities within a given higher education institution. This proposed model specifically emphasises on WIS development in order to ensure that WIS in higher education institutions has a coherent planning, implementation, and control process in place consistent with the enterprise's business strategy. The model will then be applied to support WIS development and implementation at University of Lampung (Unila) as the case study.

REFERENCES

1. R. E. Kraut, and L. A. Streeter (1995). Coordination in software development, Communications of the ACM, vol.38, no.3, pp.69-81.
2. A. W. Brown (2004). Model driven architecture: Principles and practice, Software and Systems Modeling, vol. 3, no 4, pp. 314-327.
3. J. A. Zachman (1987). A Framework for Information Systems Architecture. In: IBM Systems Journal, vol 26, no 3. IBM Publication G321-5298
4. L. G., Cretu (2012). Ontology for Semantic Description of Enterprise Architectures. In Business Information Systems Workshops (pp. 208-219). Springer Berlin Heidelberg.
5. IEEE: IEEE Standard 1471-2000 (2000). Recommended Practice for Architectural Description of Software-Intensive Systems. IEEE Computer Society, New York. Available online: http://www.iso-architecture.org/ieee-1471/ (2000)

PART I

SETTING THE STAGE

CHAPTER 1

COMPREHENSIVE MEASUREMENT FRAMEWORK FOR ENTERPRISE ARCHITECTURES

MAHESH R. DUBE and SHANTANU K. DIXIT

1.1 INTRODUCTION

American National Standards Institute/Institute of Electrical and Electronics Engineers (ANSI/ IEEE) standard 1471-2000 describes architecture as the fundamental organization of a system, embodied in its components, their relationships to each other and the environment, and the principles governing its design and evolution. Enterprise architecture is the set of representations required to describe a system or enterprise regarding its construction, maintenance and evolution. Enterprise architecture aims at creating an environment suitable for mapping the organizational assets to business processes which can identify relevance and realm of business strategy adopted. An Enterprise architecture framework typically consists of business architecture, information architecture, application system architecture, and infrastructure technology architecture.

Architecture frameworks are evaluated on the basis of scope, architecture process, verification support, standards compliance and overall complexity of the architecture. Enterprise architectures should support the

This chapter was originally published under the Creative Commons Attribution License or equivalent. Dube MR and Dixit SK. Comprehensive Measurement Framework for Enterprise Architectures. International Journal of Computer Science & Information Technology 3,4 (2011), DOI: 10.5121/ ijcsit.2011.3406 71.

business processes and indicate the benefits earned by its application. Feature extraction and enhancement are the major issues while dealing with architecture flexibility and scalability. Productivity, cost-effectiveness and optimization in terms of services are the other broad parameters affecting deployment of enterprise architectures.

It is necessary to observe the pattern of migration from platform-independent and platform-specific elements in Enterprise architecture evolution. In this paper, we are limiting the scope of views and its correspondence to The Open Group Architecture Framework (TOGAF), Generalized Enterprise Reference Architecture and Methodology (GERAM), IEEE Std 1471- 2000 IEEE Recommended Practice for Architectural Description, Model-Driven Architecture (MDA) and ISO RM-ODP.

1.2 RELATED WORK

John Zachman developed a framework in 1987 which was based on plan-driven approach and best practices adoption that can be deployed within the development organizations to address enterprise engineering problems. It was based on maintaining information profile of function aspects as well as the management required to accomplish the development activities. The prime issue addressed by Zachman's framework was architecture integration and implementation with a well-designed organization structure. Cap Gemini Ernst & Young developed an approach for analysis and development of enterprise and project-level architectures known as the Integrated Architecture Framework (IAF). IAF was the first implementation of enterprise engineering solutions which was widely accepted by technical community. Similar to Zachman's framework, IAF also aims at partitioning the problem in to manageable pieces based on the area of concern. IAF starts at Business Management aspect primarily dealing with business process and taskforce management. It maps the technology problem to information as knowledge-base, Information System used for traceability, and Technology Infrastructure, with special emphasis on Security aspects and Governance. Enterprise Architecture Planning (EAP) defines a process that emphasizes techniques for organizing and directing enterprise architecture projects, obtaining stakeholder commitment, presenting the plan to stakeholders, and leading the organization through the transition from planning to implementation [1].

Federal Enterprise Architecture Framework (FEAF) was developed in 1998 with the vision of integrating federal architectural segments. The FEAF was based on knowledge and asset management across the organization with a uniform terminology used for architectural integration. The business-driven aspect of FEAF was designed in view of accommodating the current as well as future business needs. The business information was later used in planning and implementation business operations in order to realize the Enterprise Architecture. FEAF emphasized on Architecture Evolution management with the help of transitional and transformational processes [8].

The Open Group Architecture Framework (TOGAF) was based on Application Lifecycle Management which largely covered the areas of governance as applicable to related areas of problems spanning from data to security. TOGAF is considered to be as a major contribution for enterprise architecture development because of the flexibility offered as well as verificationvalidation support provided. The Open Group is a vendor-neutral and technology-neutral consortium seeking to enable access to integrated information, within and among enterprises, based on open standards and global interoperability [2].

The IFAC/IFIP Task Force on Architectures for Enterprise Integration developed an overall definition of a generalized architecture which focused on modeling and tools that can be used for enterprise development. Generalised Enterprise Reference Architecture and Methodology (GERAM) addressed the issue of single enterprise development as well as networked enterprise development through various views which can be used at various levels of details depending on area of specialization of the enterprise. GERAM was based on Entity oriented strategy used for enterprise development [12].

The Object Management Group (OMG) introduced the Model-Driven Architecture (MDA) initiative as an approach for system development based on specification and interoperability expressed in terms of formal models. In MDA, Platform-Independent Models (PIMs) are used to represent the target system analysis and design expressed in a general-purpose modeling language, such as Unified Modeling Language (UML). The platform-independent model can be mapped to a Platform-Specific Model (PSM) by mapping the PIM to some implementation language using set of transformational rules. The MDA considers Metamodeling as a key concept for artifact generation at all stages evolution. MDA support evolution with the help of consistent mapping of resources at source to target with the help of metamodel at the two ends as well as transformation rules along with model merging [13].

The OMG MDA comprises CWM, UML, MOF and XMI as standards for model-driven development. The Common Warehouse Metamodel (CWM) defines a metamodel representing both the business and technical metadata which can be found in the data warehousing and business analysis domains. It is used as the basis for interchanging instances of metadata between heterogeneous, multi-vendor software systems. UML, which is a general purpose modelling language provides support for modelling structural and behavioural properties of the system and is part of CWM. UML is an integrated effort of three object-oriented methods (Booch, OMT, and OOSE). UML has extensive support for modelling generic systems. UML 2.0 is widely used in reactive systems behaviour analysis. The Meta Object Facility (MOF) is an OMG standard defining a common, abstract language for the specification of metamodels. It defines the four-level structure used to represent the details of how the notation repository can be made available to the modeller on model space. MOF semantics defines metadata repository that support model construction. It has the support for applying the transformations based on metamodel level selected. XML Metadata Interchange (XMI) defines XML tags that can be used to represent objects and their associations [3] [4].

1.3 ZACHMAN FRAMEWORK

The Zachman Framework for Enterprise Architecture is a widely used and accepted approach for developing or documenting an enterprise-wide architecture. It is based on Information System Architecture (ISA) and typically used in a development environment which supports organization structures and practices [5]. It is considered to be the basis for the emergence of other eminent enterprise architectures. ZF's key goals are for enterprise architecture analysis and modelling and it is concerned with perspectives of constructing an information system. The Zachman Framework organized as a table as indicated in Table 1. The rows are as follows:

- Scope: It is an executive summary for a planner.
- Business model: It indicates the business process engineering efforts and activities planned in order to achieve business goals.
- System model: It indicates data elements and software functions that represent the business model.
- Technology model: It describes the constraints of tools, technology, and materials.
- Components: It indicates smallest pieces of system that can found to be functional, tested and verified according to specification.
- Working system: It depicts the operational system.

The columns are as follows:

- Who: Represents the individuals who have enactment of fulfilment of some service.
- When: Represents achievement of explicitly stated goals or objectives by the individuals on a time line indicating activity arrival and exit.
- Why: Describes the motivations of the enterprise.
- What: Describes the activities involved in corresponding area of the enterprise.
- How: Shows the functions within each perspective.
- Where: Shows locations and interconnections within the enterprise

1.4 ISO RM-ODP

The Reference Model of Open Distributed Processing (ISO-RM-ODP) provides a framework for the development of systems that supports processing under heterogeneous platforms [6]. To model distributed systems, Object-modeling approach is used in RM-ODP. RM-ODP is a joint effort by the International Organization for Standardization (ISO), the International Electrotechnical Commission (IEC) and the Telecommunication Standardization Sector (ITUT). The problem-solution pairing can be done by the "viewpoints" which provide a way of describing the system; and the "transparencies" that identify specific problems unique to distributed systems as indicated in Figure 1. [7].

TABLE 1: Zachman Framework

Data (what)	Function (how)	Network (where)	People (who)	Time (when)	Motivation (why)	
Scope (Planner)	List of things important to business	List of processes the business performs	List of locations where business operates	List of organisations/ agents that are important	List of significant events	List of business goals/ strategies
Enterprise Model (Owner)	Semantic Model	Business Process Model	Business Logistic System	Work Flow Model	Master Schedule	Business Plan
System Model (Designer)	Logical Data Model	Application Architecture	Distributed System Architecture	Human Interface Architecture	Processing Structure	Business Rules
Technology Model (Builder)	Physical Data Model	Systems Design	Technology Architecture	Presentation Architecture	Control Structure	Rule Design
Components (Subcontractor)	Data Definition	Program	Network Architecture	Security Architecture	Timing Definition	Rule Specification

RM-ODP consists of four basic International Standards:

- Overview: It describes the overview of the ODP, Scope and terminology involved in overall architecture development.
- Foundations: It describes the significant issues and factors which should be considered for distributed processing functions and systems. .
- Architecture: It represents the characteristics possessed by distributed processing system under constraints mentioned in specification. It also recommends the use of viewpoints that can be used for logical grouping of related areas of the enterprise.
- Architectural Semantics: It focuses on the modelling with the help of formal specification techniques with adequate details of each concerned area.

The viewpoints in RM-ODP are:

- Enterprise viewpoint: It deals with the strategy that can be used to accomplish the business goals and needs as identified in the preliminary phase of problem investigation.
- Information viewpoint: It focuses on information structure, information flow, logical and physical organization of information with information change tracking.

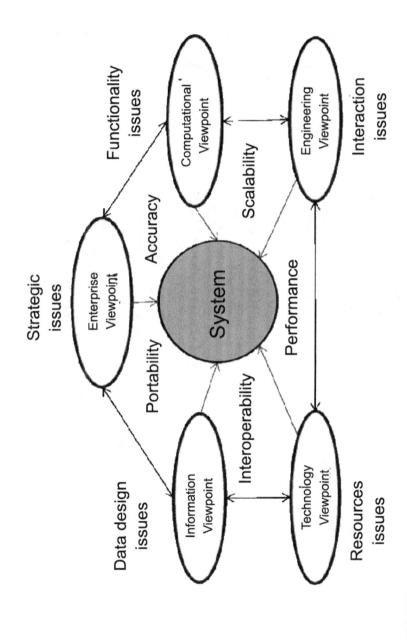

FIGURE 1: Viewpoints in RM-ODP

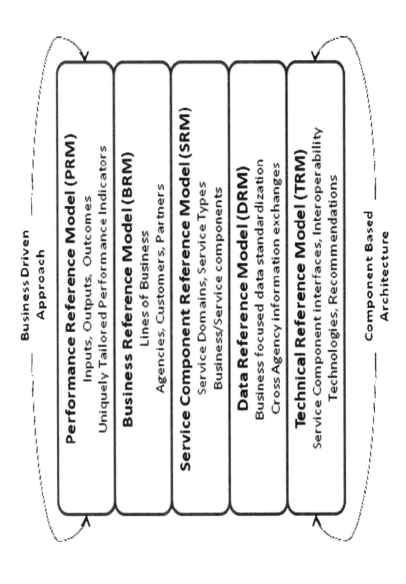

FIGURE 2: FEAF Reference Models

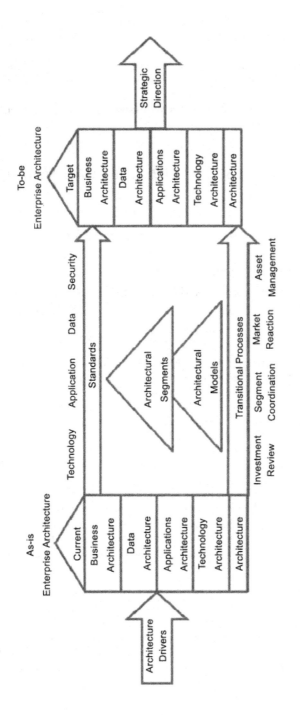

FIGURE 3: Simplified FEAF structure.

- Computational viewpoint: It focuses on structural elements of the system and their dynamics guided by protocols represented by interfaces and functionality by objects.
- Engineering viewpoint: It indicates overall organization of the objects identified and their participation in various interaction patterns to satisfy a service.
- Technology viewpoint: It indicates hardware and software components that formulate the system.

1.5 FEDERAL ENTERPRISE ARCHITECTURE FRAMEWORK (FEAF)

The goal of FEA is to improve interoperability within U.S. government agencies by creating single enterprise architecture for the entire federal government [8].

The intent of the FEAF is to enable the federal government to define and align its business functions and supporting IT systems through a common set of reference models. Figure 2 indicates FEAF Reference Models which are defined as follows:

- Performance Reference Model (PRM): The PRM is a standardized framework to measure the economics of investments and adherence to program portfolios in future based on performance.
- Business Reference Model (BRM): The BRM is a function-driven framework for describing business operations of the federal government independent of the agencies that perform them.
- Service Component Reference Model (SRM): The SRM is a framework which supports enactment of service-component relationship on the basis of performance objectives.
- Data Reference Model (DRM): The DRM is a generic model which describes the information necessary to trace operation level details.
- Technical Reference Model (TRM): The TRM is a technical framework which verifies and validates the components capabilities in relation to the specification stated and acceptable performance with reference to standards agreed upon.

The major components of the FEAF are (Figure 3):

- Architecture Drivers: It indicates the factors and conditions due to which the business scenario or target design can change over a time period.
- Strategic Direction: It consists of the vision and strategic information regarding objectives to be achieved by the target architecture. The strategic

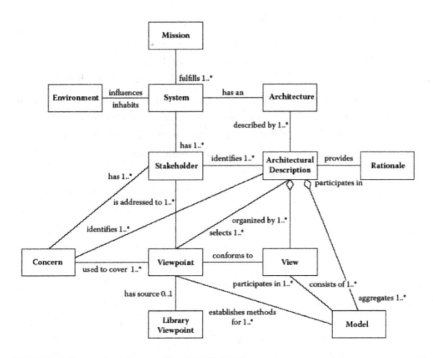

FIGURE 4: Conceptual framework of IEEE 1471

direction becomes necessary to have a pilot estimate of operational effort required to realize the enterprise solution.

- Current Architecture: It defines the "as is" scenario of the enterprise architecture and consists of existing solutions to the problem identified. It describes the capabilities needed to be addressed in accordance with the limitations of the existing solution.
- Target Architecture: It defines the "to-be-built" scenario of the enterprise architecture and consists of improved architecture and performance. It indicates the changed business needs which are required to be fulfilled in accordance with the technology migration. The target architecture can be assessed by using performance metrics indicating adherence to specification.
- Transitional Processes: It supports the migration from the current to the target architecture.
- Architectural Segments: It consists of focused architecture efforts on major crosscutting business areas.
- Architectural Models: It indicates both strategic and technical models that guide the enterprise solution which is feasible with formal representations.
- Standards: It refers to all standards, guidelines, and best practices.

1.6 IEEE1471-2000 STANDARD

The IEEE Recommended Practice for Architectural Description of Soft-ware-Intensive Systems (IEEE Std 1471-2000 aka ANSI/IEEE Std 1471-2000) introduces a conceptual model that integrates mission, environment, system architecture, architecture description, rationale, stakeholders, concerns, viewpoints, library viewpoint, views, and architectural models facilitating the expression, communication, evaluation, and comparison of architectures in a consistent manner [9].

Stakeholders are the one who are materially benefited from the solution development. The stakeholders have specific concerns and roles which should be carefully accounted while initiating and terminating the development activities. The customers or users may not have a complete view of acceptability of the solution. Therefore it is crucial to identify the stakeholder needs before the development can commence. A view indicates group of concerns as identified through partitioning of the system. A viewpoint defines a specific case of view related to a key aspect. A viewpoint indicates possible alternatives that can be considered while analyzing and designing the system rationally using appropriate modelling techniques [10]. The conceptual framework of IEEE 1471 is shown in Figure 4 and described as follows:

- A system has architecture.
- Architecture is described by one or more architecture descriptions.
- An architecture description is composed of one or more of stakeholders, concerns, viewpoints, views, and models.
- A stakeholder has one or more concerns.
- A concern has one or more stakeholders.
- A viewpoint indicates possible alternatives for relevant stakeholders.
- A view conforms to one viewpoint.
- A viewpoint defines the reason for existence of the model.
- A view can have collective representations guiding more than one view.
- A viewpoint library is composed of viewpoints.

1.7 THE OPEN GROUP ARCHITECTURE FRAMEWORK (TOGAF)

TOGAF enables corporate architects and stakeholders to design, evaluate, and build flexible enterprise architecture for the organization. The initial versions of TOGAF were based on the Technical Architecture Framework for Information Management (TAFIM), developed by the U.S. Department of Defense (DoD) [11]. There are four types of architectures that are commonly accepted as subsets of overall enterprise architecture, all of which TOGAF is designed to support:

- Business (or business process) architecture: It defines the organization structure, business processes as well as governance.
- Applications architecture: It indicates the base architecture which includes architectural segments along with their interrelationships that conforms to business processes of the organization.
- Data architecture: It describes the data management capabilities grouped to logical as well as physical assets supporting application realization.
- Technology architecture: It is concerned with the infrastructural capabilities which should be considered while implementing and deploying the enterprise solution. As platform independence is a prime issue to be dealt in service composition and availability, it describes the technological alternatives available to male system resources available.

TOGAF has following views and viewpoints for development of enterprise. As mentioned previously, this may be regarded as taxonomy of viewpoints by those organizations that have adopted ANSI/IEEE Std 1471-2000.

- Business Architecture Views, which address the concerns of the users of the system, and describe the flows of business information between people and business processes
- Data Architecture Views, which address the concerns of database designers and database administrators, while identifying and normalizing the database entities of the system.
- Applications Architecture Views, which address the concerns of system and integration engineers responsible for developing and integrating the software components of the system.
- Technology Architecture Views, which address the concerns of acquiring the commercial off-the-shelf (COTS) components that may reduce the cost of software development. The amendments to the components falls into white-box and black-box modifications made to the components. It depends on the suitability of the existing components to identified services to be realized.

1.8 GENERALIZED ENTERPRISE REFERENCE ARCHITECTURE & METHODOLOGY (GERAM)

Previous research, carried out by the AMICE Consortium on CIMOSA, by the GRAI Laboratory on GRAI and GIM, and by the Purdue Consortium on PERA, has produced reference architectures which were meant to be organizing all enterprise integration knowledge and serve as a guide in enterprise integration programs. The IFIP/IFAC Task Force concluded that the architecture derivation should have unique purpose and satisfy the service demands and business needs with a possibility of retainment of service capabilities of previous reference architectures. The recognition of the need to define a generalized architecture is the outcome of the work of the Task Force [12].

The GERA life-cycle for any enterprise consists of different life-cycle phases that define types of activities that are pertinent during the life of the entity. Life-cycle activities encompass activities that span from identification to realization of the enterprise or entity. The activities can be broken into lower level tasks in order to manage the operational effort. Traditional lifecycle management is evident in GERAM methodology with a shift from process components to entities.

- Entity Identification: It describes the entities that constitute the enterprise problem and their limits with possible interactions within the system as well as the external environment. This can be treated as scoping of entities identified..
- Entity Concept: It deals with entity's mission, vision, values, strategies, objectives, operational concepts, policies, business plans which can be used to create entity's knowledge base for further development processes initiation.
- Entity Requirement: The activities needed to develop descriptions of operational requirements of the enterprise entity, its relevant processes and the collection of all their functional, behavioral, informational and capability needs.
- Entity Design: It indicates the process of solution structure and specification of individual components that conforms to the requirements specified.
- Entity Implementation: It describes the effort needed to implement the components identified during the Entity Design step. Reusable components can also be used in concern with cost of modification. If cost of modification of components is higher, components from scratch can be implemented.

- Entity Operation: It deals with deployment of product or service at the customer end. It deals with transition of the solution from source environment to target environment with identification of problems at customer end while using product or services.
- Entity Decommissioning: These activities are needed for future issues like refactoring, reengineering problems associated with the product or services. It emphasizes on the new demands raised to reconsider the problem due to training or design issues.

1.8.1 MODELING FRAMEWORK OF GERA

GERA provides an analysis and modeling framework which is based on the life-cycle approach and indicates following dimensions for defining the scope and content of enterprise modeling.

- Life-Cycle Dimension: providing for the controlled modeling process of enterprise entities according to the life-cycle activities.
- Genericity Dimension: providing for the controlled particularization (instantiation) process from generic and partial to particular.
- View Dimension: providing for the controlled visualization of specific views of the enterprise entity.

1.8.1.1 ENTITY MODEL CONTENT VIEWS

Four different model content views define for the user oriented process representation of the enterprise entity descriptions The Function View represents the functions contained in individual business processes and the control applied to each one of them at operational level. The Information View formulates the knowledge base about the entities and the objects identified so as to address the mission and objectives of the enterprise. The Resource View represents hardware, software and human resources required to realize the enterprise solution. The Organization View represents the roles and responsibilities of the people concerned with enterprise development. It also deals with the accountability of human resources in the organization.

1.8.1.2 ENTITY PURPOSE VIEWS

- The Customer Service and Product View represents the contents relevant to the enterprise entity's operation and to the operation results.
- The Management and Control View represents the contents relevant to management and control functions necessary to control that part of the enterprise entity that produces products or delivers services for the customer.

1.8.1.3 ENTITY IMPLEMENTATION VIEW

- The Human Activities View represents the set of tasks that are required to be achieved in order to realize the entities identified along with clear description of responsibilities.
- Automated Activities View is an indicator of automation effort required to be estimated and delivered to address the technological aspects. This view indicates the tasks that can be automated so as to reduce the manual processing overheads.

1.8.1.4 ENTITY PHYSICAL MANIFESTATION VIEWS

- The Software View represents all information resources capable of controlling the execution of the operational tasks in the enterprise
- The Hardware View represents the physical resources that are needed to achieve the product functionalities or services at the source and target environments of the enterprise.

1.9 MODEL DRIVEN ARCHITECTURE (MDA)

Model Driven Architecture was introduced by Object Management Group to allow long-term flexibility of implementation, integration, and testing of products and services. Interoperability and platform independence were the two major concerns addressed by MDA. MDA was significantly different approach for specification-based modeling of systems which concentrated on models as a prime issue than objects as in case of object oriented methodologies. MDA introduced model composition and transformation from three levels of models i.e. from Computation-Independent

Model (CIM) to Platform-Independent Model (PIM) to Platform-Dependent Model (PSM) based on mapping rules [13]. The core technologies of the OMG MDA are the UML modeling language, the Meta Object Facility (MOF) and the Common Warehouse Metamodel (CWM). Organization of a software system can be represented by structural elements or classes with their interfaces that comprise or form a system and behavior represented by collaboration among these elements. UML is not associated to a process model since it supports the engineering activities ranging from requirements to realization. MOF provides the basis for defining metamodels and model repositories. CWM provides the baseline for data warehousing and data integration. Models are formal specifications of system. A formal specification is consists of syntax, semantics for constructs formulation and usage [14]. The models of the system fall into following categories:

- The conceptual model that captures the system in terms of the domain entities that exist and their association with other system environments.
- The logical view of a system that captures the abstractions indicating the logical separation and boundaries of each identified entity in the conceptual model. It also describes the mechanism through which these entities will interact and form realizable behaviour.
- The physical model of a system describes the software and hardware components that form the system solution space conforming to the specification.

A model can exhibit static structure and defines the universe of discourse. It requires concept mapping from the application domain to a well-formed structure. The analysis classes are transformed to design classes and later to software classes with implementation details of interaction pattern amongst the objects [15]. Dynamic behaviour can be modelled as the life history of one object as it interacts with the rest of the world; the other is

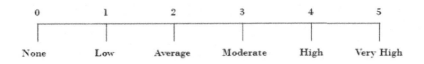

FIGURE 5: Measurement Scale

the communication patterns of a set of connected objects as they interact to implement behaviour or as the view of an object in isolation is a state machine, a view of an object as it responds to events based on its current state, performs actions as part of its response, and transitions to a new state. Following are the views of "4+1" view architecture:

- Use Case view: It focuses on scenarios indicating the functional require-ments which will be used by external entities. This view incorporates analy-sis level information that dictates static behaviour of the system along with further decomposition of the functionalities.
- Design view: It represents the logical structures which support the require-ments expressed in the case view described in terms of classes (and objects) and their behaviour (including interactions between them). It encompasses classes, interfaces, and collaborations that define the vocabulary of a sys-tem and supports functional requirements of the system.
- Implementation view: It incorporates physical components that can be grouped into packages indicating realized entities. The basis for these com-ponents is analysis and design level classes. The class hierarchy and inter-action profile are preserved in this view.
- Process view: It deals with dynamic interaction profile of object including concurrency, time and flow of control. Process view is important in case of real-time applications where synchronization is an important dimension.
- Deployment view: It consists of executables in the form of nodes. De-ployment view indicates the resources of system in implementation environment.

The Model-Driven Architecture consists of CIM, PIM, and PSM indi-cating how they should be used in context of system generation. A view-point indicates an aspect or concern of the system which is identified using abstraction principles. A viewpoint model or view of a system is a repre-sentation of the domain or partition under consideration. The details of a view can help organize the system elements into realizable components. A platform is a set of functionalities relevant to technology indicating avail-ability of usable services and resources. The platform independence can be achieved by hiding the details of service profiles at software architecture level from the application level by introducing interfaces which can make the resource available from one platform to the other.

TABLE 2: Comparison by higher order goals

Comparison Parameter	ZF	RMODP	FEAF	TOGAF	IEEE 1471	MDA	GERAM
Architecture Definition & Understanding	3	5	5	5	5	5	5
Architecture Process	0	0	5	5	5	5	5
Architecture Evolution Support	0	3	5	5	5	5	5
Architecture Analysis	5	5	5	5	5	5	5
Architecture Models	5	5	5	5	5	5	5
Architecture Knowledge Base	0	5	5	5	5	5	5
Abstraction	4	3	4	4	3	4	5
Application Architecture	3	2	3	3	3	4	4
Architecture Continuum	4	4	4	4	3	4	4
Architecture Governance	3	3	4	4	3	4	3
Architecture Landscape	3	3	3	3	3	4	3
Architecture Verifiability	0	3	0	5	5	5	5
Baseline	4	2	3	4	4	3	3
Business Governance	4	3	3	5	3	4	4
Capability Architecture	3	4	3	4	3	4	3
Data Architecture	3	2	4	4	2	3	3
Design Tradeoffs	3	3	4	3	4	3	3
Design Rationale	3	5	4	5	5	5	5
Data Governance	3	2	4	4	2	3	3
Enterprise Continuum	3	4	4	5	3	4	4
Environment Management	4	3	3	4	3	4	3
Foundation Architecture	1	3	2	5	4	4	4
Gap Analysis	3	3	3	5	4	3	4
Metamodel	3	2	2	5	4	4	4
Performance Management	2	2	2	4	2	4	4
Standardization	0	5	3	5	5	5	5
Total	69	84	92	115	98	108	106

- Computation Independent Viewpoint: The computation independent viewpoint focuses on requirements of the system and its structure with environmental needs. It indicates customer, user and stakeholder's perspectives and expectations from system.
- Platform Independent Viewpoint: The platform independent viewpoint focuses on analysis and design models of the system which incorporates the system elements identified and their explicit relationships without adherence to implementation details.
- Platform Specific Viewpoint: The platform specific viewpoint indicates implementation level details of the system elements specific to a particular platform. This can be accomplished by using mapping and transformation rules for migrating from PIM to PSM.

1.10 MEASUREMENT PROCESS

Measurement is the process of describing entities in terms of numbers or symbols. It also indicates the uniqueness property that should be preserved by each identified entity [Fenton 95]. Thus, measurement requires entities (objects of interest), attributes (characteristics of entities) and rules (and scales) for assigning values to the attributes. Measures and metrics are based on measurement scales which can be derived from the rules that we use for assigning values to attributes. Different rules lead to different scales. An ordinal scale permits measured results to be placed in ascending (or descending) order. However, distances between locations on the scale have no meaning. We have used ordinal scale having score values ranging from 0 to 5 as indicated in Figure 5.

1.10.1 COMPARISON BY HIGHER ORDER GOALS

Enterprise integration begins with identification of mission and objectives that directs the business needs of the customer. The enterprise problem then broken into domains that can be implemented and integrated so as to form the enterprise segments [16] [17]. The success of enterprise acceptance depends on customer needs realization and its fulfillment. All enterprises follow a life cycle from their initial concept through a series of stages or phases comprising their development, design, construction, operation and maintenance, refurbishment or obsolescence, and final disposal. Table 2 indicates comparison by higher order goals [18]. Following list indicates higher order goals for Enterprise Architecture:

- Architecture Definition and Understanding—it describes the terminology and guidelines that must be used to define the architecture framework conforming to the needs as stated by the stakeholders identified.
- Architecture Process—it the set of activities performed to attain architecture construction.
- Architecture Evolution Support—it maintains traceability and change profile of system evolution.
- Architecture Analysis—it is a process used to determine the aspects, view and viewpoints that makes up basis of architecture segments.
- Architecture Models—it represents the system in terms of analysis and design models that conforms to standards and specification that guides the development plan.
- Architecture Knowledge Base—it maintains the information base of significant design decisions that directs the enterprise architecture rationale.
- Abstraction—it is an approach to classify the system elements based on similarities and differences. It leads to identification of unique entities of the system.
- Application Architecture—it describes the logical entities and components along with their interaction pattern conforming to identified business needs.
- Architecture Continuum—it is an information base that keeps records of identified architectural segments with appropriate and adequate details so as to realize the architecture. It also encompasses strategies and reference model dictating adoption of architectural styles.
- Architecture Governance—it is the set of processes that guides management and control of the enterprise architectures and other issues related to enterprise-wide level development.
- Architecture Landscape—it deals with identification and management of enterprise assets in accordance with stakeholder needs. It indicates the processes and plans which incorporates strategic and operational profile of the enterprise conforming to stakeholder needs.
- Architecture Verifiability—it provides the set of properties and characteristics that can be checked in order to review the service or product functions.
- Baseline—it is a specification indicating agreed upon properties and characteristics of system that can be examined with current deliverables to estimate its performance. It also serves as important dimension in addressing the changes to be incorporated and its control.
- Business Governance—it indicates the business processes, policies and regulations that need to be practiced while developing the enterprise.
- Capability Architecture—it indicates specification of architectural components with detailed implementation and compositional semantics.

- Data Architecture—it describes the data resources grouped into logical and physical compartments guiding organizational assets.
- Design Tradeoffs—it offers the alternatives for selecting rational design from available choices in order to address the diverse business and technical needs.
- Design Rationale—it indicates the proof of statements for verification and review decisions.
- Data Governance—it indicates the verification mechanisms used to ensure that the data properties and structure has adequate support for transformation and migration.
- Enterprise Continuum—it describes the process of classification of architecture segments and components that makes up the enterprise. It also maintains the catalogue of reference models used; foundation architectures referred leading to custom architectures.
- Environment Management—it indicates the source and target environment in which the system will be operational. It describes the set of resources, facilities and information base that should be made available to deploy enterprise solution.
- Foundation Architecture—it is an architecture of generic services and functions that provides a base for construction of architectural components in question.
- Gap Analysis—it is an indicator of differences between two representations. It is performed to estimate acceptance level of enterprise architecture designed and the baseline considered.
- Metamodel—it is model about model. It specifies the detailed structure and semantics of architectural properties specifications.
- Performance Management—it indicates the post-development activities that needed to be followed to keep track of application performance after deployment.
- Standardization—it indicates whether the determined and accepted standards are met or not.

TABLE 3: Architecture Definition and Understanding

Score	Factors indicating Degree of influence
0	Enterprise Scope and focus is not defined.
1	The extent of enterprise and architectural effort required to attain the same is defined.
2	A complete architecture domain description consisting domain information with resource and time constraints is specified.
3	The level of detail of architecture and architecture effort is determined.
4	Timing considerations for Architecture Vision realization are indicated.
5	Target Architecture and Transition Architecture alternatives are defined in order to address the stakeholder objectives in order with increments.

TABLE 4: Architecture Process

Score	Factors indicating Degree of influence
0	Organizational context for conducting enterprise architecture is not defined.
1	Organizational context for conducting enterprise architecture is defined and reviewed.
2	The sponsor stakeholder(s) and other major stakeholders impacted by the business directive are identified to create enterprise architecture and determine their requirements and priorities.
3	The elements of the enterprise organizations affected by the business directive are identified and scoped with constraints and assumptions.
4	The framework and detailed methodologies to be used for developing enterprise architectures in the organization concerned are defined.
5	Target Architecture, infrastructure and supporting tools are selected and implemented.

TABLE 5: Architecture Analysis

Score	Factors indicating Degree of influence
0	The life cycle management principles and commitments are not defined; hence realistic schedule of architecture development is absent.
1	Preliminary phases of life cycle are defined and the overall realm of architecture framework is defined and formally stated
2	The Key Process Areas (KPA) as well as the Key Performance Indicators (KPI) are defined with adherence to the corresponding business processes and drivers .
3	The Baseline Architecture effort with the relevant stakeholders, and their concerns and objectives is defined.
4	The development schedule and performance metrics to meet are developed.
5	Formal approval plan and impact analysis of development cycle is established.

TABLE 6: Architecture Verifiability

Score	Factors indicating Degree of influence
0	No Architecture Verification iteration exists.
1	Architecture Context iterations indicating architecture approach, principles, scope, and vision is established.
2	The iterations required to establish correct and stable architectural information base is established and revised with relevant technical drivers.
3	Transition Planning iterations supporting formal change adoptions for a defined architecture is established.
4	Architecture Governance iterations supporting governance of change activity progressing towards a defined Target Architecture is established.
5	The opportunities and migration planning are traced.

TABLE 7: Architecture Governance

Score	Factors indicating Degree of influence
0	Governance principles are not established and hence no architecture verification can exist.
1	All the stakeholders of the enterprise development have agreed upon the processes and deliverables as stated by the stakeholders and recorded by the organization.
2	All actions implemented and their decision support is available for inspection by authorized organization and provider parties.
3	All processes, decision-making, and mechanisms used are established so as to minimize or avoid potential conflicts of interest.
4	Performance metrics and practices to be followed to ensure the architecture enactment policies are determined and monitored.
5	Stakeholder participation and interaction is determined to monitor progress and performance of architecture development. It principally yields the client and development organization neutral scenario to deploy architectural solution successfully.

Table 8: Business Governance

Score	Factors indicating Degree of influence
0	No description of the Baseline Business Architecture.
1	Major domain areas and architectural elements are identified formulating the productlevel-functions and services. Target architecture scope and applicability in corresponding environment are determined.
2	Reviews of Target Business Architectures and baselines are conducted and examined.
3	Architecture views and viewpoints are established in accordance with the stakeholder needs and concerns in order to reveal stable architecture segments.
4	Organization, Goals, Role and Business Service catalogue is developed and standards for each building block from reference model are selected.
5	Cross check of overall architecture and Architecture Repository mapping is performed.

TABLE 9: Standardization

Score	Factors indicating Degree of influence
0	Enterprise architecture program is not defined.
1	The enterprise architecture processes and standards are derived by ad hoc means and are not formal enough to guide the business strategies.
2	The vision and mission of target enterprise architecture is established with stable and explicit business strategies.
3	The architecture is well defined and communicated to human resources and management with operation details and responsibilities assigned. It also covers the initial investments to be made along with procurement processes and control.
4	Enterprise architecture documentation is maintained so as to control and trace the changes incorporated in ongoing development cycle.
5	Metrics and measures are established and practiced to verify the architecture process. The areas for improvement and optimization of business processes are identified.

Table 3 to Table 9 indicates the selection criteria on the measurement scale 0 to 5. Architecture Governance, Business Governance and Standardization are the key parameters which determine the applicability of the enterprise architectures depending on the business domain and context identified. Architecture Process and Verification are the other parameters which can be useful in adjudging suitability of the enterprise architecture at the construction and deployment stages. Architecture Analysis depends on baselines and Key Performance Indicators (KPIs).

1.10.2 COMPARISON BY NFR SUPPORT

Requirements are a specification of functions or services that should be accomplished by the system. The requirements are the properties and characteristics possessed by the system along with satisfaction of constraints on them. Requirements vary in intent and in the kinds of properties they represent in terms of product parameters and process parameters. Product parameters are can be further classified as functional requirements (FR) which indicate what the system should do and affects the performance of the system directly whereas non-functional requirements (NFR) indicate what the system should do and affects the performance of the system indirectly [19] [20].

NFRs are particularly difficult to handle and tend to vary significantly if the goals are expressed ambiguously. Many non-functional requirements have emergent properties. Such requirements cannot be addressed by a single component, but depends for their satisfaction on how all the system components inter-operate. Correctness, consistency, traceability and requirement interaction management are the prime issues to be dealt [21]. Unfortunately, non-functional requirements may be difficult to verify. Non-functional requirements should be quantified. If a non-functional requirement is only expressed qualitatively, it should be further analyzed until it is possible to express it quantitatively. The non-functional requirements mentioned below are quantified on the scale as indicated in the measurement process. Table 10 indicates the comparison by NFR support. Following are the NFRs considered:

- Cohesiveness—It is the degree to which each module in a system does one task and does it well. Cohesion refers to the uniqueness of purpose of the system elements.

- Conceptuality—It represents the concepts in the domain under study. With a conceptual perspective, developers may conceive of what the customer requires, not how. The conceptual level is more abstract than the implementation level, in which the details of how the requirement is to be met are manifested in the code itself.
- Configurability—It describes the ability to organize and control elements of the software configuration. A system's software configuration is defined as the items that comprise all information produced as part of the software process.
- Consistency—It describes two aspects of a system's design and development. Consistency may refer to the use of approaches and techniques describing the system specifications which leads to uniform representations of the system.
- Coupling—It describes the degree to which the modules and components of a given system rely on and interact with other modules and components of that system.
- Diversity—It describes the degree of difference between a system's components and modules. It refers to the degree of difference between data structures and data types throughout a program.
- Extensibility—It involves extending both the design of the system and the software system itself. It describes the degree to which architectural, data, or procedural design can be extended by adding variations to an already stated theme.
- Standardizability—It indicates acceptability and conformance of deliverables against standards. The process standard defines the procedures or operations used in making or achieving a product; the product standard defines what constitutes completeness and acceptability of items that are produced as a result of a process.
- Adaptability—It is defined by the rate at which the software solution can adapt to a new requirement. Adaptability also refers to the degree to which a system may be changed based on a pre-existing system or an unalterable constraint.
- Dependability—It describes the degree to which software performs expected functions and services without failure and acceptable precision.
- Flexibility—It describes the effort required to modify an operational program or system. A software system may be required to be flexible if there will be known a change in its operating environment after it has been deployed and is in normal operation.
- Maintainability—It describes the effort required to locate and fix an error in a program. It the ease with which a program can be corrected if an error is encountered, adapted if its environment changes, or enhanced if the customer desires a change in requirements.
- Maturity—It describes the degree to which a software system is mature. A system is said to be mature when it has attained a final, desired state of full development.
- Portability—It describes the ease with which the software can be transposed from one environment to another.
- Scalability—It refers to the ease with which a system may be made smaller or larger, although most of the time, increasing the system's size is the concern, not reducing it.

- Robustness—It describes the degree to which a program or system can recover gracefully whenever a failure occurs. It also describes the time it takes the system to restart after experiencing system failure.
- Security—It describes the mechanisms that detect the possible threats to programs and data. It may also refer to the probability that the attack of a specific type will be repelled.
- Compatibility—It describes the ability of two or more systems to exchange information. When a system is being deployed to replace an earlier version of that system, it is imperative that it be compatible with everything that it is replacing is compatible with.
- Inter-operability—It is defined as the ability of the systems to exchange the services with agreed protocols and architectural support at both the ends.
- Usability—It describes the effort required to learn and handle the services or product functions over a period of time.

TABLE 10: Comparison by NFR support

Comparison Parameter	ZF	RMODP	FEAF	TOGAF	IEEE 1471	MDA	GERAM
Adaptability	4	4	3	5	4	5	4
Compatibility	3	4	3	5	3	4	4
Cohesiveness	3	3	4	4	4	4	4
Conceptuality	4	4	4	5	4	4	4
Configurability	2	4	4	4	4	4	4
Consistency	3	3	4	5	4	4	4
Coupling	3	3	4	5	4	4	4
Diversity	3	3	3	5	3	5	3
Dependability	3	4	4	4	4	4	4
Extensibility	3	3	4	4	3	4	4
Flexibility	3	4	3	5	4	4	4
Inter-operability	3	3	3	5	3	5	3
Maintainability	3	4	4	4	3	4	3
Maturity	3	3	3	4	4	4	4
Portability	2	4	3	4	3	4	3
Robustness	3	4	4	4	3	4	4
Scalability	3	3	4	4	4	4	4
Security	2	3	4	4	3	4	3
Standardizability	3	3	4	5	4	4	3
Usability	4	3	3	5	3	4	3
Total	60	69	72	90	71	83	73

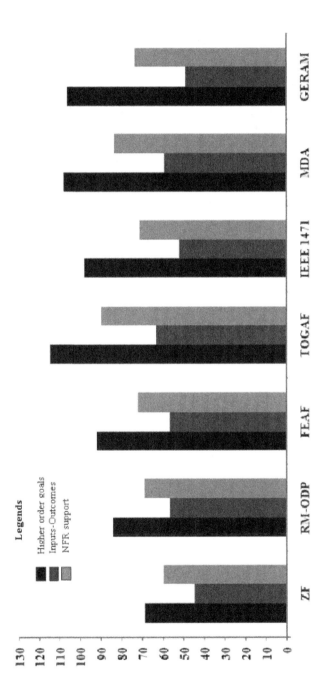

FIGURE 6: Consolidated Comparison Chart

1.10.3 COMPARISON BY INPUTS AND OUTCOMES

Business drivers, Technology inputs, and Business requirements focus on the problem issues in view of the stakeholders. The context and relevance of the problem scenario can be further broken into various model supports as indicated in Table 11. The process enablers as well as process measures are key areas determining sustainability and stability of the enterprise solution.

TABLE 11: Comparison by Inputs and Outcomes

Comparison Parameter	ZF	RMODP	FEAF	TOGAF	IEEE 1471	MDA	GERAM
Business Drivers	3	3	5	5	3	5	3
Technology Inputs	0	3	5	5	5	5	4
Business Requirements	5	5	5	5	3	5	3
Information System Environment	3	5	5	5	5	4	4
Existing Architecture Evaluation	3	5	5	5	5	4	5
Business Model Support	5	5	5	5	3	5	3
System Model Support	5	5	5	5	5	5	4
Information Model Support	5	5	5	5	5	4	4
Computation Model Support	5	5	5	5	5	5	4
Software Configuration Management	0	3	0	5	4	4	4
Software Process Incorporation	4	4	4	5	3	4	3
Implementation Model	3	4	4	4	3	4	4
Platform	4	5	4	4	3	5	4
Total	45	57	57	63	52	59	49

Figure 6 indicates the consolidated chart representing the enterprise architecture suitability depending on higher order goals, NFR support and input-outcomes.

1.11 CONCLUSIONS

The paper covers a broad discussion of major enterprise architecture methodologies. The enterprises can be categorized into small-sized, medium-sized and large-sized enterprises depending on the range of problem issues, business requirements, and organization portfolio. It is significantly difficult to decide on selecting a specific enterprise architecture methodology due to the changes that drives the enhancement scenario for these methodologies. Every system development effort is constrained by the time, scope and cost triplet. The relationship between scope and performance has to be established at the time of system conceptualization so that realistic solution with required fitness criteria can be developed.

The paper proposed an ordinal scale based measurement process for measuring enterprise architecture methodologies in terms of higher order goals, NFR support and input-outcomes. It can be observed that TOGAF and MDA are the most successful methodologies in addressing the issues indicated due to incorporation of views and viewpoints. Business, Architecture, Technology and Data governance are also the key areas which indicate the rationale and applicability of the methodologies. However, the fundamental methodology proposed by Zachman Framework is nearly adopted and considered by every descendant methodology development effort.

The paper focused on the criticality of addressing NFR issues. NFR properties are the abilities that the system should possess that ensure required quality and performance has been met at product or service level. We have considered major NFRs that can impact the selection of enterprise architecture methodologies. It can be observed that TOGAF, MDA, GERAM and IEEE 1472-2000 are in a comparable range in this context. The paper also suggests that there cannot be a radical shift from one methodology to the other since methodology mapping must be discovered before doing so. Finally, the selection of any enterprise architecture methodology will depend on organization culture, mission, principal investment at the initial phase and adherence to the architecture principles.

REFERENCES

1. L. Bass, P. Clements & R. Kazman, (1997) Software Architecture in Practice, Addison-Wesley, Reading, MA.
2. Bachmann F., Bass L., Klein M. & Shelton C., (2005) Designing software architectures to achieve quality attribute requirements, IEE Proceedings – Software 152 (4), 153–165.
3. D. Chen & F. Vernadat, (2004) "Standards on enterprise integration and engineering—state of the art," International Journal of CIM, Vol. 17, No. 3, pp. 235–253.
4. Jeff A. Estefan, (2007) Survey of Model-Based Systems Engineering (MBSE) Methodologies, INCOSE MBSE Focus Group, pp 1-47
5. J. A. Zachman, (1987) "A framework for information systems architecture," IBM System Journal, vol. 26, no. 3, pp. 276–292.
6. ISO/ITU-T (1997), Reference Model for Open Distributed Processing (ISO/ITU-T 10746 Part 1 - 4), Information Standards Organization.
7. K. Farooqui, L. Logrippo, and J. de Meer, "Introduction into the ODP Reference Model," 2/14/96, Department of Computer Science, University of Ottawa, Ottawa K1N 6N5, Canada; Research Institute for Open Communication Systems Berlin (GMD-FOKUS), D10623 Berlin, Hardenbergplatz 2, Germany
8. CIO-Council, (1999) Federal Enterprise Architecture Framework version 1.1, URL http://www.cio.gov/archive/fedarch1.pdf, Accessed 21 December 2010.
9. P. Clements, (2005) "1471 (IEEE Recommended Practice for Architectural Description of Software-Intensive Systems)," CMU/SEI-2005-TN-017, Software Architecture Technology Initiative, Carnegie-Mellon Software Engineering Institute
10. IEEE Architecture Working Group, (2000) "IEEE Recommended Practice for Architectural Description of Software-Intensive Systems, IEEE Std 1471-2000," IEEE, Tech. Rep.
11. The Open Group, (2003) The Open Group Architecture Framework (ver 8.1 Enterprise Edition), URL http://www.opengroup.org/architecture/togaf/#download, Accessed 21 December 2010.
12. IFIP-IFAC Task Force, (1999) "GERAM: Generalized Enterprise Reference Architecture & Methodology," IFIP-IFAC Task Force on Architectures for Enterprise Integration, Tech. Rep.
13. Reda Bendraou, Philippe Desfray, Marie-Pierre Gervais & Alexis Muller, (2008) MDA Tool Components: a proposal for packaging know-how in model driven development, Journal of Software and System Model Vol3: pp 329–343, Springer-Verlag.
14. Kruchten P, (1995) The 4+1 View Model of Architecture, IEEE Software, 12, 6 pp 42-50.
15. Jacobson I., Booch G. & Rumbaugh J., (1999) The Unified Software Development Process, Addison-Wesley.

16. Dobrica L. & Niemela E., (2002) A survey on software architecture analysis methods, IEEE Transactions on Software Engineering 28 (7).

17. Nelly Condori-Fernández & Oscar Pastor, (2008) Analyzing the Influence of Certain Factors on the Acceptance of a Model-based Measurement Procedure in Practice: An Empirical Study, MODELS'08 Workshop ESMDE, pp 61-70

18. Benoit Vanderose & Naji Habra, (2008) Towards a generic framework for empirical studies of Model-Driven Engineering, MODELS'08 Workshop ESMDE, pp 71-80

19. Chung L. & Nixon B., (1995) Dealing with Non functional Requirements: Three Experimental Studies of a Process-Oriented Approach, Proceedings of the 17th International Conference on Software Engineering, pp. 24–28.

20. Cysneiros L. & Leite J., (2004) Non functional Requirements: From Elicitation to Conceptual Models, IEEE Transactions on Software Engineering 30 (5), 328–350.

21. Sadana V. & Liu X., (2007) Analysis of conflicts among non-functional requirements using integrated analysis of functional and non-functional requirements. Computer Software and Applications Conference, COMPSAC.

CHAPTER 2

EVALUATION OF ARIS AND ZACHMAN FRAMEWORKS AS ENTERPRISE ARCHITECTURES

MELITA KOZINA

2.1 INTRODUCTION

Structured frameworks called enterprise architectures capture and manage the complexity of modern organizations. Today, modern organizations are based on connectivity of the business system model and its relevant information system (IS) model. Such integral business systems are highly complex systems consisting of elements such as objectives, data, people, processes, technology. These systems require coordination and integration in order to manage the existing interdependencies between all these elements.

Enterprise architecture frameworks present a conceptual map necessary for building an integral business model supported by the relevant IS. It requires identification information about the organization from different perspectives (views). The main perspectives are: data, functions, networks, organizational structures, schedules and strategy. Each of there business perspectives gradually get implemented into the components of a future IS. This process is supported by the second dimension of the conceptual map containing certain phases (levels). Consistently, frameworks support various management tasks such as business process improvement,

This chapter was originally published under the Creative Commons License or equivalent. Kozina M. Evaluation of ARIS and Zachman Frameworks as Enterprise Architectures. Journal of Information and Organizational Sciences **30**, 1 (2006).

workflow management, software engineering and in general, business system management.

Building enterprise architectures involves modelling techniques in order to capture the organization in its entirety, an adequate methodology to establish a basis for business system management, and a lifecycle concept of IS as well as integrated modelling tool in order to build and maintain these architectures. This paper analyzes two important frameworks for enterprise architecture: the Architecture of Integrated Information Systems (ARIS) and the Zachman framework. The goal of this paper is to compare these approaches according to defined criteria, to determine the points of their complement and to assess how efficient this approach is to business system management. In order to solve this, in the paper we describe the general phases of business system management (Chapter 2), analyze, compare and evaluate ARIS and Zachman frameworks (Chapters 3, 4) and define the generic model of business system management supported by said architectures (Chapter 5).

2.2 BUSINESS SYSTEM MANAGEMENT

One of the characteristics of the business system management is continuous optimization of business processes and relevant IS implementation. The following process changes could become necessary for business process optimization (business process improvement): changing the process structure; changing organizational structure and responsibilities; changing the data in use; discussing possible outsourcing measures (shifting from internal to external output creation); implementing new production and IT resources to improve processing functions. Figure 1 shows the procedural model of business system management. Through the entire management, we can recognize two main levels: business level and IS development/ implementation level.

The starting point of the entire process is the strategic plan and the goals of the business system, for which an adequate IS is developed. Business applications need to support the realization of business goals, which is why information needs stem from the business technology model.

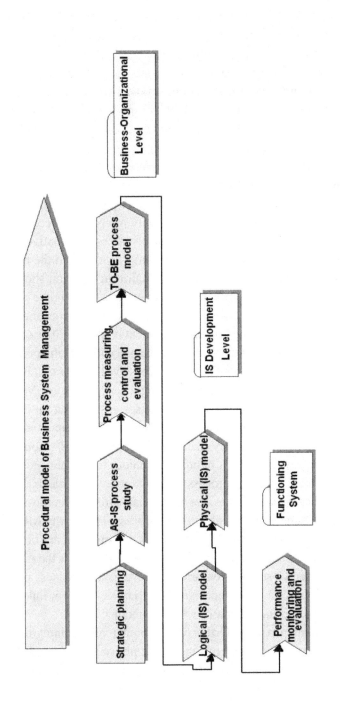

FIGURE 1: Procedural model of Business System Management

Business process design (AS-IS model)/ redesign (TO-BE model) aims to achieve the greatest efficiency possible in terms of business-organizational solutions. When engineering optimal business processes, different models (diagram techniques) can be provided using ARIS/Zachman architectures and adequate integrated tools and methods. This is the only way to get a picture of the existing process and all its aspects (AS-IS model).

The reference models can be also included, along with available knowledge on best practices. It is also possible to compare alternative procedures (benchmarking) or carry out simulation studies. Multiple alternatives are generated, studied and analyzed in simulation studies in order to engineer the best possible business process. Defining and analyzing the various engineering alternatives in "what-if" situations are also necessary. In dynamic simulations the dynamic behaviour of process alternatives is studied. Furthermore, ICT innovations can also have an effect on the new design of business processes. For example, if the firm in question should implement standard integrated software, the business model would be customized using standard reference models. Also available are various methods for ensuring the quality of the processes (like ISO9000).

In order to design business processes in accordance with specific goals, they must be evaluated in accordance with process goals. Data obtained from lower management levels ensure the basis for process evaluation. Measuring, controlling and evaluating business processes ensure data on efficiency of business processes, resource utilization, process throughput times, process qualities and process costs. By changing process priorities, resource allocations and processing sequences in order to achieve the process goals, process owners can manipulate the process itself. In order to be able to plan and control current business processes, the appropriate information must be made available to the persons responsible for the process. This is why ARIS/ Zachman concepts and compatible integrated tools and methods are being used. For example, an efficient method for cost structure analysis is the ABC method (Activity Based Costing). Measuring and evaluating business processes visualize various reasons for redesign (TO BE model) and for linking design and controlling/evaluating phases with one another.

The next important phase in further development of the integrated business system is the development of logical and physical models of the future IS, i.e. gradual transformation of business aspects into IS compo-

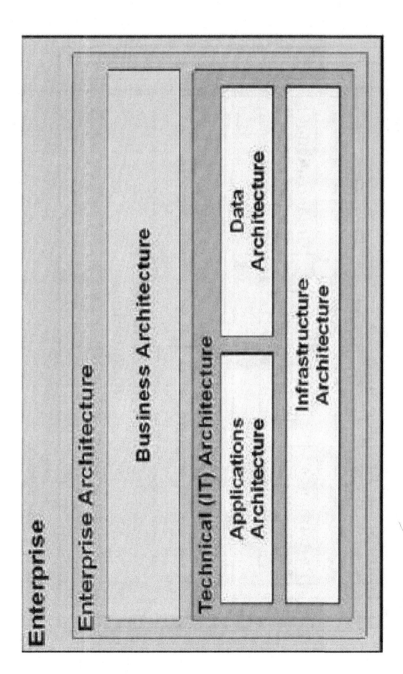

FIGURE 2: The classic enterprise architecture

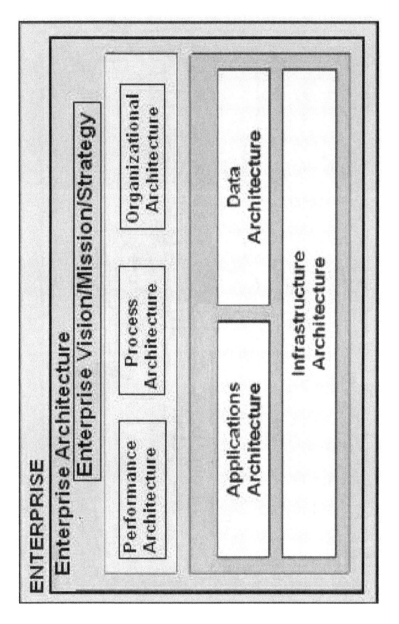

FIGURE 3: The balanced enterprise architecture

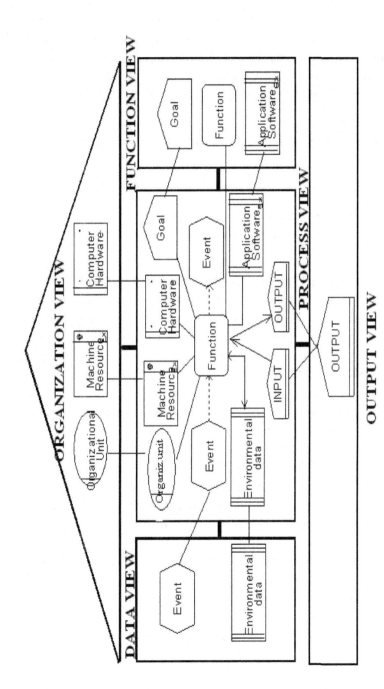

FIGURE 4: Views of the ARIS business process meta-model

nents. So constructed integrated business system demands continuous performance measuring and evaluation aimed at checking the efficacy of business applications in the realization of business goals. Furthermore, it is a good base for further AS-IS studies and new optimization of business processes (TO BE model).

2.3 ENTERPRISE ARCHITECTURE FRAMEWORKS

2.3.1 EVOLUTION OF ENTERPRISE ARCHITECTURE

The concept of enterprise-wide information technology architecture was developed very quickly. The concept included Application, Data and Technology Architectures. As a result, the enterprise architecture became software oriented and application development oriented. However, it was still not enough, since a good alignment between business processes and requests, i.e. needs, as the starting point in the design of IT solutions, could not be established. Therefore, to ensure effective mappings and requirements tracing, business architecture was incorporated into the concept.

The merging of business and IT concepts was a big step forward in the evolution of the enterprise architecture. Although the developed architectures were still too technically-oriented, the alignment between the business needs and IT solutions grew none the less. Business elements such as functions, processes, organizational units and users were introduced to ease the task of software requirements analysis and to increase the quality of application development results. Fig. 2 shows a classic case of enterprise architecture.

Further development found the emergence of comprehensive IT oriented enterprise architecture comprised of Business Process, Organizational and Business Performance Architectures (performance metrics on strategic, operational and human levels). There are many benefits of such architectures such as [3]:

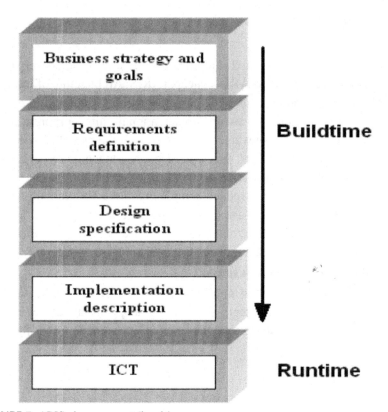

FIGURE 5: ARIS phase concept (levels)

- Faster response to environmental threat or opportunity;
- Better alignment of change initiatives with enterprise strategy;
- Better alignment of application systems with business objectives and needs;
- Reduced application development lifecycles;
- Reduced application maintenance costs;
- Improved operating procedures.

Fig. 3 shows the balanced enterprise architecture concept.

2.3.2 THE ARIS FRAMEWORK

Business process model aimed at modelling and optimization is very complex. With that in mind, the ARIS concept (ARIS: Architecture of Integrated Information Systems) allows us to reference the following important business aspects (meta-classes) [4]:organizational units; corporate goals; initial and result events; messages; functions; material output, service output and information services; financial resources; machine resources and computer hardware; application software; human output; process environmental data. In order to reduce complexity, meta-classes with similar semantic interrelationships are grouped into ARIS views [4, 7, 9]. The grouping of classes and their relationships into views serves the purpose of structuring business process models. Figure 4 shows five different views.

FUNCTION VIEW -HOW, WHY

The processes transforming input into output are grouped in a function view. Due to the fact that functions support goals, goals are also allocated to function view. In application software, computer-aided processing rules

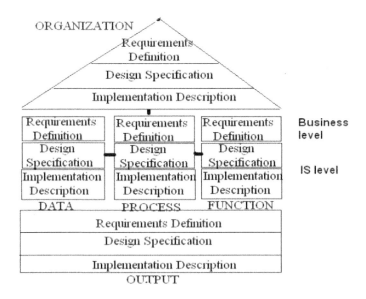

FIGURE 6: The ARIS architecture

of a function are defined. Thus, application software is closely aligned with functions, and is also allocated to function view.

ORGANIZATION VIEW – WHO, WHERE

The class of organization view creates the hierarchical organization structure. This view is created in order to group responsible entities (department, team, position, person, role) or devices executing the same work object.

TABLE 1: ARIS modelling techniques (ARIS Toolset)

View	Description Level of System Lifecycle	ARIS Modelling Techniques
Organization	Requirements Definition	Organizational chart Shift calendar Location allocation diagram
	Design Specification	Network topology
	Implementation	Network diagram Technical resource model
Data	Requirements Definition	Authorization hierarchy CD Diagram Cost category diagram DTD, DW-structure eERM, SAP structured ER model eERM attribute allocation diagram Material diagram Information carrier diagram IE data model Technical terms model Data warehouse
	Design Specification	Attribute allocation diagram Relations diagram System attributes System attribute domain
	Implementation	Table diagram
Function	Requirements Definition	Function tree Objective diagram SAP ALE function model SAP applications diagram Y diagram
	Design Specification	Application system type diagram
	Implementation	Application system diagram

TABLE 1: *Cont.*

Control/Process	Requirements Definition	Authorization map
		Business framework
		Business control diagram
		c3 method; class diagram
		Classification diagram
		Communications diagram
		Competition model
		DW-transformation
		E-business scenario diagram
		Extended event driven process chain (eEPC)
		Event diagram
		Function allocation diagram
		Function/organizational level diagram
		Industrial process
		Information flow diagram
		Input/output diagram
		Knowledge structure diagram
		Material flow diagram
		Office process
		Privileges diagram
		Process selection diagram, Process matrix
		Product/Service exchange diagram
		Role diagram, rule diagram
		SAP ALE models
		Screen design
		UML Models
		Value-added chain diagram
	Design Specification	Access diagram
		Program flow-chart
		Program structure chart
		Screen diagram
		Data flow diagram
	Implementation	Hardware/software allocation diagram

DATA VIEW – WHAT, WHEN

This view comprises the data processing environment as well as the messages triggering functions or being triggered by functions.

OUTPUT VIEW -WHAT

Output view contains all physical and non-physical input and output, including funds flows.

ARIS PROCESS VIEW- WHAT, HOW, WERE, WHO, WHEN, WHY:

Relationships among the all views, as well as the entire business process, are modelled and documented in this view.

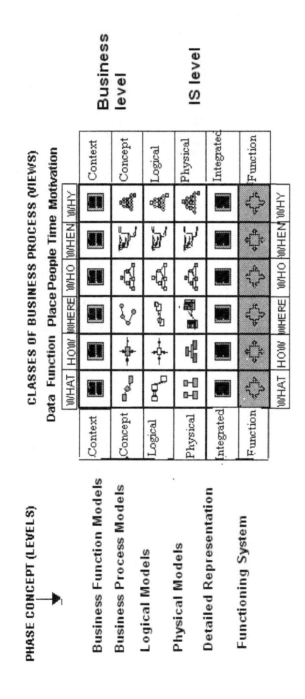

FIGURE 7: The Zachman architecture

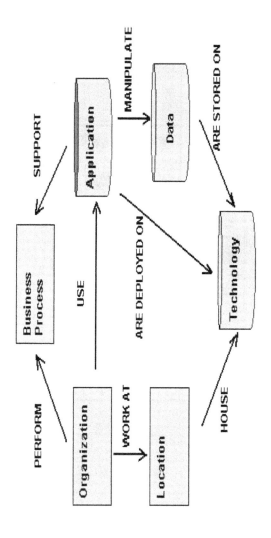

FIGURE 8: Main business aspects within the model of enterprise architecture

TABLE 2: Characteristics of the Zachman concept in the building enterprise architecture

Views Levels	Data (what)	Process (How)	Place (Where)	People (Who)	Time (When)	Motivation (Why)
Business function modelling (Enterprise areas)	High-level data classes related to each function	High-level business functions (Enterprise areas)	Locations related to each function	Stakeholders related to each function	Events related to each function	Business objectives and performance measures related to each function
Business process models	Business data	Business processes (Business Dynamics Model)	Locations related to each process	Organization roles and responsibilities in process	Events related to each process	Policies, procedures, and standards for each process
Logical models	Logical data models	Logical representation of IS (System Dynamics Model)	Logical network model	Logical representation of access privileges constrained by roles	Logical events and their triggered responses constrained by business events and their responses	Policies, procedures, and standards associated with a business role model
Physical models	DBMS type requirements constrained by logical data models	Specifications of applications that operate on particular technology platforms (Function Dynamics Model)	Specification of network devices and their relationships within physical boundaries	Specification access privileges to specific platforms and technologies	Specification of triggers to respond to system events on specific platforms and technologies	Business rules constrained by IS standards
As built	Data definition constrained by technology platforms	Programs coded to operate on specific technology platforms	Network devices configured to conform to node specifications	Access privileges coded to control access to specific platforms and technologies	Timing definitions coded to sequence activities on specific platforms and technologies	Business rules constrained by specific technology standards
Functioning enterprise	Data values stored in actual databases	Functioning computer instructions	Sending and receiving messages	Personnel and key stakeholders working within their roles and responsibilities	Timing definitions operating to sequence activities	Operating characteristics of specific technologies

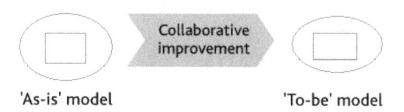

'As-is' model   'To-be' model

FIGURE 9: Business Process Improvement

In addition to the five views (vertical dimension), ARIS involves a concept of different description levels (horizontal dimension). As it is presented in Figure 5, these five levels represent design aspects according to the well-established software engineering lifecycle.

The source point of this concept is the business strategy and strategic goals (level 1) defined in accordance with the vision and mission of the business system. Goals and global business processes (enterprise areas), resources and potential effects of IS implementation supported by new ICT technologies, are determined during the phase of strategic planning. The second phase, requirements definition (level 2), is related to business process modelling and ICT requirements specification (generally $\Rightarrow$ databases, application software, hardware, networks, etc.) This is the most important phase of integral system development, since the model of business technology represents the starting point in IS development and its implementation. Thanks to developed methods (especially ARIS and Casewise Corporate Modeler), it is possible to use different diagrams for business process modelling. This is especially important considering the fact that the business process is a complex unit and it is difficult to show all its aspects using one model (diagram). The third phase, design specification (level 3), features the beginnings of gradual transformation of business descriptions into ICT objects. The logical IS model is developed and ICT components described in more detail (database models, hardware and network specifications, software modules, etc.) Actual IT products are still irrelevant. Phase 4, implementation description (level 4), contains descriptions of the technological model of the future system, and business requirements are implemented into actual hardware and software components. These phases of IS development are known as "buildtime" phases. Subsequently, the completed system becomes operable, meaning it is followed by an operations phase, known as "runtime".

Thus, this dimension establishes the physical linkage from business strategy to information technology through translation from a strategic to an operational level. The implementation of business models in information systems represents a modern approach in the development of integral business systems. The ARIS architecture accomplishes this by linking its business aspects (ARIS VIEWS) with the phase concept (ARIS LEVELS). In this way, the views of meta-business process model are valid for all descriptive levels. Figure 6 shows the ARIS architecture of an integral business system. This ARIS "House" illustrates the 15 component of this framework.

2.3.2.1 ARIS MODELLING TECHNIQUES

The ARIS framework is supported by the tool of the same name and the methodology containing various modelling techniques. These modelling techniques (and their relationship with additional techniques) enable the development, analysis and optimization of enterprise architecture. In table 1, adequate modelling techniques for each ARIS component are recommended.

ARIS provides an extensive number of model types to build complex enterprise architectures. These model types represent different modelling methods. Though each model type has its own modelling objects (entity types, classes, functions or objectives), all objects are integrated in one comprehensive meta-model. This meta-model is the conceptual foundation of ARIS and inter-relates each and every model type.

2.3.3 THE ZACHMAN FRAMEWORK

The Zachman concept is another framework for modelling, evaluation, optimization, management and documenting of the integral business system.

Figure 7 shows the Zachman architecture of the integral business system [1, 10]. Table 2 shows basic characteristics of the Zachman architecture with respect to development and implementation of the business model into the IS model.

Meta-classes of the business process (aspects of business modelling) are also divided into individual views in the Zachman architecture in order to reduce the business model complexity. Main description views in Zachman's concept are: data-process-place-people-time-motivation. The phase concept of this architecture includes a number of levels in relation to the ARIS architecture. Buildtime lifecycle in the Zachman architecture contains the following phases: business function modelling-level 1; business process models- level 2; logical models (IS model)- level 3; physical models (IS model)-level 4; "As Built" IS models-level 5. Runtime phase is related to functioning enterprise-level 6.

2.4 COMPARISON AND EVALUATION OF ARIS AND ZACHMAN FRAMEWORKS - FINDING THEIR COMPLEMENTARITY

In this chapter we compare the ARIS and the Zachman frameworks and determine their complementarities. The following criteria are defined in order to accomplish this:

  a) Consistent framework for modelling, analyzing and optimizing the enterprise architecture;
  b) Two dimensional structure;

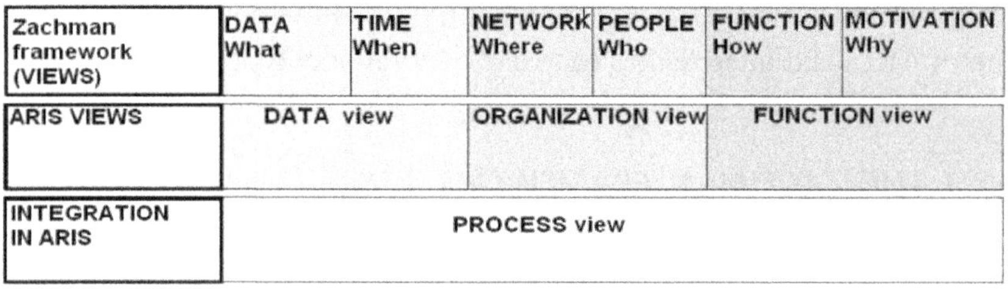

Zachman framework (VIEWS)	DATA What	TIME When	NETWORK Where	PEOPLE Who	FUNCTION How	MOTIVATION Why
ARIS VIEWS	DATA view		ORGANIZATION view		FUNCTION view	
INTEGRATION IN ARIS	PROCESS view					

FIGURE 10: Vertical comparison and evaluation of ARIS/Zachman frameworks

c) Drill-down approach;

d) Multi-user repository;

e) Adequacy of using ARIS modelling techniques within Zachman's framework.

2.4.1 CONSISTENT FRAMEWORK FOR MODELLING, ANALYZING AND OPTIMIZING THE ENTERPRISE ARCHITECTURE

ARIS/Zachman frameworks supported by adequate integrated tools and methods enable improvement teams to build a complete visual model of the enterprise architecture (address the organization in its entirety). Main business aspects within the model of enterprise architecture are business processes, the people that perform them, the location where they occur, applications that support the business processes and manipulate relevant data and used technology (the IT hardware and network). These business aspects are shown on Figure 8.

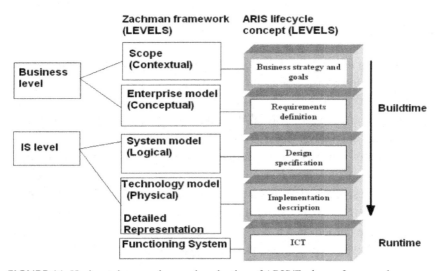

FIGURE 11: Horizontal comparison and evaluation of ARIS/Zachman frameworks

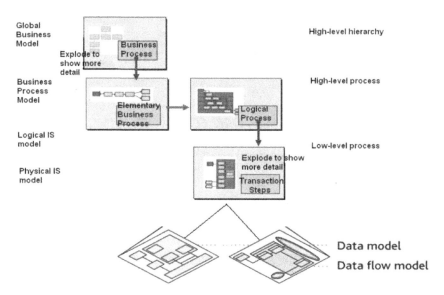

FIGURE 12: Drill down approach

In co-relation with the above, the ARIS/Zachman frameworks supported by integrated tools and methods link together the business process, organizational aspects, IT architecture and data modelling through a multiuser repository. Organizational/hierarchical modelling: form a top-down view of the organization within the scope of improvement initiatives; structures organizational entities and their roles in the realization of business processes.

Business process modelling/analysis/simulation [2,4]: visualize and improve business processes using easy-to-understand modelling notation; ensure jobs flow through departments, systems, suppliers from initial customer enquiry through to final delivery of product/service; analyze the "AS-IS" model, design the "TOBE" model (shown on Figure 9); to accomplish this, they use various assessment procedures, reference models, and of special importance, simulation methods, which research the dynamics of business processes in order to locate inefficiencies and quantify the benefits of "what-if" scenarios.

There are numerous reasons which point to the importance of creating models of business processes. Here are some of them:

	DATA VIEW		ORGANIZATION VIEW		FUNCTION VIEW		
	DATA (What)	TIME (When)	PLACE (Where)	PEOPLE (Who)	FUNCTION (How)	MOTIVATION (Why)	
Business Function Models (Context)	Business framework						Business level
Business Process Models (Concept)	Data warehouse Material diagram eERM		Organizational Chart Location allocation diagram		Function tree Objective diagram		
Logical Models Physical Models	Relations diagram System attributes		Network topology		Application system type diagram		IS level
	Table diagram		Technical resource model		Application system diagram		
Detailed Represent							
Functioning System	data	schedule	network	organization	application		

FIGURE 13: ARIS modelling techniques within Zachman's framework

- To get a picture of the existing process and all its aspects (business process meta-classes),
- To measure, monitor and evaluate business processes (quality, expenses, time, flexibility, allocation of resources, redundancy, "what-if" assessment, etc.),
- To make an adequate optimization of the business process in question using various methods (reference models, best practices, simulation, benchmarking)
- To structure the ISO 9001:2000 quality management system,
- To allow for the development of own integrated IS or to customize standard software solutions.

IT Architecture modelling: optimizes how the business applications, hardware, networks and data structures align to support or automate the business processes; design where and how to integrate or replace existing systems to maximize ROI (return on investment).

Data flow modelling: analyzes how information is shared across an organization to perform business processes; create a hierarchical set of data flow diagrams (DFDs) to show data flows at every level of operations.

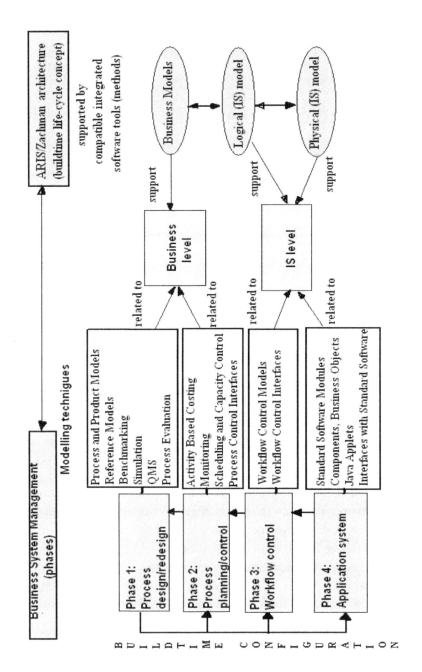

FIGURE 14. Business system management supported by ARIS/Zachman architectures

Data modelling: analyzes and improves the use of data at each step of business process; implement data design or analyze how existing data structures support the business.

2.4.2 TWO-DIMENSIONAL STRUCTURE

Both approaches have a similar design paradigm, as they both include a twodimensional framework. The six vertical dimensions (descriptive views) in Zachman's framework are analogous to the views in the ARIS concept. Thus, both frameworks decrease the complexity of modelling the business technology and facilitate modelling within the following perspectives: what, how, where, who, when, why. In addition, the six horizontal dimensions (descriptive levels) in Zachman's framework are analogous to the levels (software lifecycle concept) in the ARIS concept. Their vital importance is to connect the complete business model with the IS model.

2.4.2.1 VIEWS – REDUCTION OF COMPLEXITY IN MODELLING BUSINESS TECHNOLOGY

According to Zachman's framework, we can distinct 6 aspect of business technology: data, time, network, people, function and motivation. The ARIS concept is primarily focused on business processes, data, functions and organization (shown on Fig. 4). Data and functions correspond directly with the Zachman's Framework. The time aspect of Zachman's framework is implicitly part of the data view in ARIS concept, because the modelling object "event" represents stages of business objects in terms of time and logical sequence. In this context, messages trigger the functions or being triggered by functions. The organization view in ARIS concept captures both, people and location (network) on different levels of modelling. In addition, objective diagrams within the function view in ARIS cover the motivation field in Zachman's framework. All vertical description views of Zachman's framework can be found in ARIS framework. Determined complementarities of ARIS and Zachman frameworks according to this criterion are shown on Fig. 10.

ARIS framework includes the integrative and the dynamic process view. Integrating the other views constitutes the process view. This leads to a high level of consistency of the enterprise architecture.

The four main ARIS views (organization, data, function, process) ensure that enterprise architectures can be captured in its completeness and from various aspects. These aspects as the main business drivers within the enterprise architecture are: business (processes, organization, location), applications, information and technology.

2.4.2.2 LEVELS – INTEGRATION OF BUSINESS MODEL AND IS MODEL

Descriptive levels (horizontal dimension of both frameworks) ties the business model and the IS model. In that context, we can place these levels in two categories: the business level and the IS level. According to Zachman's framework, we differentiate between 6 levels, while the ARIS framework supports 5 levels. Complementarities of ARIS and Zachman frameworks recognized according to this criterion are shown in Fig. 11.

The "Scope (contextual)" level in u Zachman's framework corresponds to the "Business strategy (business goals)" level of the u ARIS framework. Strategic goals are formed within these levels, as well as global business areas (enterprise areas), and global effects of IS implementation are estimated. The "Enterprise model (conceptual)" level in Zachman's framework corresponds to the "Requirements definition" level of ARIS, and this level is significant due to detailed modelling of business processes and planning of information needs, which are determined within the said level. Listed levels all refer to the business level. Furthermore, the development of the integrated business system is furthered by the development of the logical IS model (level 3 in both frameworks). Levels like "Technological model of the system" and "Full system integration" in Zachman's framework correspond to the "Implementation description" level of the ARIS concept. These levels refer to the IS level. Both approaches support the runtime level, "Functioning system".

Both approaches allow drilling down from a strategic to an operational level by providing a map, which covers the complete cycle from strategy to implementation of information systems. These lifecycle concepts for information systems connect the described levels (areas) and support their interrelations.

2.4.3 DRILL DOWN APPROACH

Drilling down approach enables us to navigate down from high-level overviews to low-level processes, so we can present architecture at the appropriate level of detail.

Multi-level process modelling allows us to develop process models from the global hierarchy structure; first in the higher level models (business level), then in the lower levels (IS level). Each level of process modelling is supported by the relevant data model and the model of technological architecture. This can be observed in most detail on the physical level of IS. Fig. 12 shows the drill down approach supported by the ARIS/Zachman frameworks.

In addition, drill across approach enables us to link from the end of one business process to the start of the next.

2.4.4 MULTI-USER REPOSITORY

Both frameworks (ARIS and Zachman) are developed so that they can incorporate integrating tools and modelling techniques which are grouped in a single, unique data base—the multi-user repository. When we create the business object representing a certain business aspect (process, organization, data, application, IT architecture), we save its main attributes in the data base (capture key business knowledge). That same object may be used by any member of the project team in any model that the object appears in (reuse objects). This provides a coherent approach to business improvement, where all users are working with a common set of objects representing the enterprise. The functionality also induces reuse of best-practice processes across the organization.

Due to the possibility related to reusing objects throughout models, keeping models up-to-date is quick and easy. We can update an object in any one part of a model, for all other occurrences of that object to instantly be updated. In this way, we can identify all the areas of the organization that would be impacted by changing one component of the business.

2.4.5 ARIS MODELLING TECHNIQUES CAN BE UTILIZED WITHIN ZACHMAN'S FRAMEWORK

Of great importance for more effective enterprise architecture management are the methods and the tools supporting the concepts, i.e. frameworks for enterprise architecture. The ARIS framework is supported by the said methods and integrating tools. There are methods and tools supporting the Zachman's framework as well (for example, the Corporate Modeler/Casewise). Since the goal of this paper is to determine certain points of complement between the ARIS and Zachman frameworks, this Chapter deals with the possibilities of using ARIS modelling techniques within Zachman's framework. ARIS is able to capture a complete picture of data, applications, organizations and processes based on integrated modelling techniques and tools. The ARIS concept can be regarded as complementing Zachman's concept and captures its dimensions in a very similar way. According to the two-spanned congruent dimensions of the ARIS and the Zachman frameworks, it is possible to utilize ARIS modelling techniques within the Zachman framework. Figure 13 shows how some of ARIS modelling techniques (as shown in Table 1) can be suitable within Zachman's framework. ARIS' main advantage lies in the integrative process/control view (integrating all modelling objects of the other views). Overall, both approaches are highly complementary.

2.5 BUSINESS SYSTEM MANAGEMENT SUPPORTED BY ARIS AND ZACHMAN FRAMEWORKS

The goal of the paper comes through in this chapter as defining the general phases (levels) of the business system management and examining the applicability of ARIS and Zachman frameworks as complementary architectures to these levels. General process management system phases may be described through the following levels [4,5,11]: at level 1 (process design/redesign); at level 2 (process planning and control); at level 3 (workflow control); at level 4 (application system). Figure 14 shows the global model for business system management supported by ARIS/Zachman architectures.

Complementary frameworks for enterprise architectures such as ARIS and Zachman frameworks support the business system management from organizational engineering to IT implementation, including continuous process improvement. This all becomes possible through the use of integrated software tools and methods. Phases of a business system management may be described/listed as follows:

Process design/redesign: business processes are modelled using various modelling techniques, and they cover all aspects of business modelling; various methods for optimizing, evaluating and ensuring the quality of the processes are also available.

Process planning and control: current business processes are planned and controlled by means of different methods for scheduling and capacity, cost analysis methods (activity based costing (ABC)), and other methods for quality control and business process success.

Workflow control: converts business processes into IT tools. Generally, it is not possible to administer an entire business process with one application software system. Often, a variety of systems for sales, purchasing, manufacturing, accounting is necessary. It therefore makes sense to allocate the responsibility for comprehensive process control to an explicit system level rather than distributing it across several systems. This phase is called "workflow". Workflow systems pass the documents to be processed from one work place to the next. In general, business technology flows are made up of data exchanged by business processes and the defined source and destination of their exchange; business processes are just key points in the transformation of information and data. At this phase, ARIS/Zachman architectures and compatible workflow modelling techniques provide prototyping functionality and interfaces with different workflow systems.

Application systems: documents delivered to the workplaces are specifically processed, i.e. functions of the business process are executed using computer-aided application systems-ranging from simple processing systems to complex standard software solution modules-business objects and java applets. Traditional standard software solutions are transaction driven, integrated business application systems. The main idea of componentware is to assemble software systems from individual standard components developed by various vendors. Today, new application software is generally developed using object oriented technologies. At this phase, generic business objects for logistics solutions are available, in addition to interfaces for starting standard software solutions.

Process design/redesign and process planning/control phases refer to the business level, whereas workflow control and application system phases refer to the IS level. With software installed at all levels of business system management, the ARIS/Zachman life cycle model is applicable to all levels (phases). Therefore, these architectures can be used for managing business processes from organizational engineering to IT implementation, including continuous process improvement.

Phases (levels) of business system management are interlinked by two-way communication. Using the top down approach, the model of business processes is implemented into the relevant IS (phase concept of integral business system). Accordingly, the business system model is the starting point of IS development. Using the bottom up approach, one achieves feedback from lower levels of management toward the higher ones, such as process control and process evaluation levels. Feedback is sought (and achieved) in order to facilitate new optimization and continuous improvement of business processes, i.e. the entire system (continuous system improvement). For example, process control delivers information on the efficiency of current processes. Workflow control reports actual data on the processes to be executed (amounts, times, organizational allocations) to the process control level. Each new optimization of business processes (business process reengineering) requires new business concepts which will be implemented into the relevant IS during the next (new) cycle.

2.6 CONCLUSION

Enterprise architectures provide the platform for business system management. They are an important starting point for various initiatives requiring IT support based on clearly specified business objectives. Building and managing enterprise architectures will not be successful without a stable and consistent enterprise architecture framework supported by compatible integrated tools and methods.

ARIS and Zachman enterprise architecture frameworks pave the way for engineering, planning and controlling business processes, optimization and continuous business process improvement. These concepts aid in capturing a wide range of descriptive aspects of business processes, allocating relevant methods to them. Understanding the what, how, who, where, when and the why is very important for the build-time life-cycle concept of IS. This paper analyzes the main aspects of the meta-business process for both concepts (called the ARIS/Zachman views). Views of said architectures decrease the complexity of business process modelling. However, the ARIS/Zachman architectures surpass the process architecture and enhance it using the phase concept, which starts with the business level and ends with the IS implementation level (called the ARIS/Zachman levels).

Furthermore, the paper contains the definitions of criteria for the comparison and evaluation of these two concepts, aimed at finding certain complementarities and advantages of each of the concepts. The criteria are categorized thusly: a consistent framework for modelling, analyzing and optimizing the enterprise architecture; two dimensional structure; drill-down approach; multi-user repository; and modelling techniques.

Both approaches were developed independently, but they are highly complementary. The ARIS framework is able to depict all dimensions and levels of the Zachman framework. In comparison to the Zachman framework, ARIS has a particular advantage. ARIS includes the integrative process/control view that integrates all modelling objects of the other views. This leads to a high level of consistency of the enterprise architecture.

By analyzing the details of the ARIS/Zachman architectures complement, the paper aims to offer an analysis of their applicability through the all phases (levels) of the business system management (business and IS levels). In accordance with the above, the paper further strives to define the global model for business system management supported by the ARIS/Zachman architectures. With software installed at all levels of business system management, the ARIS/Zachman life cycle model is applicable to all levels (phases). In conclusion, these architectures can be used for managing business processes from organizational engineering to IT implementation, including continuous process improvement.

REFERENCES:

1. Hokel, T.A.: The Zachman Framework for Enterprise Architecture :An Overview, Available from: http://www.frameworksoft.com/web-sitedownloads/FSI-Enterprise-Architecture-Overview-Paper.pdf, Copyright 1993-2005, Accessed: 2005-04-20.
2. Perkins, A. (1997): Enterprise Information Architecture, White Papers, Available from: http://www.ies.aust.com, Accessed: 2005-04-15.
3. IDS Scheer (2005): Enterprise Architectures and ARIS Process Platform, White Papers, Available from: http://www.changeware.net/doc/wp_ea.pdf, Accessed: 2005-12-10.
4. Scheer, A.W. (1999): ARIS – Business Process Frameworks, 3rd edition, Berlin et al.
5. Scheer, A.W. (1999): ARIS – Business Process Modeling, 3rd edition, Berlin et al.
6. Schekkerman,J. (2005): Trends in Enterprise Architecture 2005: How are Organizations Progressing?, EA Survey, Institute for Enterprise Architecture Developments, Available from: http://www.enterprise-architecture.info/Images/EA%20Survey/Enterprise%20Architecture%20Survey%202005%20IFEAD%20v10.pdf, Accessed: 2005-12-10.
7. Spiekermann, S.(2004): System Analysis and Modelling with ARIS, Available from: http://www.wiwi.hu-berlin.de/~sspiek/ITSD_VL5.ppt, Accessed: 2005-04-10.
8. Sowa,JF., Zachman J.A.(1992): Extending and formalizing the framework for information systems architectures, IBM Systems Journal, 31(3), pp.590-616.
9. Ulrich F.(2002): Multi-Perspective Enterprise Modeling (MEMO) – Conceptual Framework and Modeling Languages,Proceedings of the 35th Hawaii International Conference on System Sciences Available from http://csdl2.computer.org/comp/proceedings/hicss/2002/1435/03/14350072.pdf; Accessed: 2005-12-13.
10. Zachman,J.A. (1996): Concepts of the Framework for Enterprise Architecture, White Papers, Available from: http://www.ies.aust.com
11. Business Process Management Exchange (2004), Report, Available from: http://www.iqpc.com/binary-data/IQPC_CONFEVENT/pdf_file/6565.pdf, Accessed:2005-04-20.

CHAPTER 3

AN INTEGRATED ENTERPRISE ARCHITECTURE FRAMEWORK FOR BUSINESS-IT ALIGNMENT

NOVICA ZARVIĆ and ROEL WIERINGA

3.1 INTRODUCTION

Businesses describe their enterprise architecture using an Enterprise Architecture Framework (EAF), which is a structure for documenting the architecture of their IT systems. Usually, each business uses its own EAF, which may or may not be documented. If undocumented, the EAF is a kind of implicit conceptual metamodel of the architecture of their IT systems. However, when different businesses want to cooperate, they have to relate their EAFs to each other, and this means they should document their EAFs. While doing so a common understanding of each others EAFs is needed. We do not claim that businesses should replace the EAF they work with and that all change to the same EAF, but as far as existing EAFs are built on different abstraction mechanisms it is necessary to understand each others frameworks in order to be able to communicate in a reasonable way. An EAF, capable to integrate existing frameworks, is useful for this task, because it can show how the integrated EAFs relate to each other.

This chapter was originally published under the Creative Commons License or equivalent. Zarvic N and Wieringa RJ An Integrated Enterprise Architecture Framework for Business-IT Alignment. In: Proceedings of the 18th International Conference on Advanced Information Systems Engineering (CAiSE'06), 5-9 Jun 2006, Luxembourg *(2006).*

Very little has been written to date about EAFs. Langenberg and Wegmann [1] map the research field but make no attempt at comparing EAFs.
Schekkerman [2] lists a lot of EAFs without comparing them. Tang et al.
[3], finally, compare EAFs but do not attempt to integrate them. In this
paper we compare and integrate some of the well-known EAFs. Section 2
discusses and defines the terms Enterprise Architecture and Enterprise Architecture Framework. Section 3 presents a particular framework, called
GRAAL framework, and shows how the other frameworks relate to it.
Section 4 then presents our integrated framework and discusses the impli-
cations for business-IT alignment in value constellations.

3.2 DEFINING ENTERPRISE ARCHITECTURE AND FRAMEWORKS

We start from the concept of a system as any coherent collection of elements [4]. Software systems are systems, the set of all applications of an
organisation is a system, and organisations are systems too. We define
architecture of a system as "the structure of a system, consisting of the relationships among its components, the external properties of those components, and the way these create emergent properties with added value for
the environment." Like the IEEE architecture definition [5], we consider
there to be a single architecture of a system. There can be many different
views of an architecture [6], each of which documents a different aspect
of the architecture, but we think that there is a single structure which is
the combination of all views. Note also that the architecture of a system is
not just the structure of the system, but it includes the way in which this
structure creates an added value for the environment of the system [7].

The concept of Enterprise Architecture (EA) is defined by various
sources as the structure of the IT systems of an enterprise, or even of the
entire enterprise, or sometimes as an analysis and documentation of this
structure rather than the structure itself [8,9,10,11]. We define an EA simply by restricting ourselves to IT systems in an enterprise context: "An
enterprise architecture is the structure of an enterprise, consisting of the
relationships among its ICT systems, the external properties of those ICT
systems, and the way these create emergent properties with added value
for the enterprise." The term EAF, finally, is used mostly to indicate a list

of important abstraction mechanisms, such as perspectives, viewpoints, architectures, dimensions, etc. To be neutral with respect to the abstraction mechanisms used, we define an enterprise Architecture Framework as "a documentation structure for Enterprise Architectures." A company can use an EAF to structure descriptions of architectures in such a way that these descriptions can be compared in a meaningful way, to control the design of interfaces among IT systems, to create a repository for storage and retrieval of EA documentation, or as a set of guidelines that assists the development of an EA [12,13,11,14]. In this paper, we look at its role in connecting IT systems of different enterprises.

3.3 A COMPARISON OF EA FRAMEWORKS

The GRAAL Framework. To compare EAFs, we use one framework as reference, namely the GRAAL framework used in our architecture research [15,12,16,17]. The GRAAL (Guidelines Regarding Architecture Alignment) framework derives from a framework for information systems development methods [18,19] and has been used in the GRAAL project1 to compare architecture alignment in different organisations in the Netherlands.

The GRAAL framework divides an organisation and its IT into a number of layers, where each layer contains entities (systems in the general sense) providing services to entities in the layers above it. From the bottom up we distinguish the physical infrastructure layer, the software infrastructure layer, the enterprise system layer (i.e. applications and information systems), the enterprise, and its environment (See the examples in Fig. 1). Each company may add more detail to a particular layer, such as for example distinguish different infrastructure domains, or distinguish enterprise systems into information systems and applications. These are refinements, and the GRAAL framework contains only the greatest common divisor of all these company-specific frameworks. The essential characteristic of the GRAAL layers is that entities at one layer provide services to entities at higher layers.

The second GRAAL dimension is that systems at each level have certain aspects. Foremost among these is that each system provides services (to systems in higher layers); each service should provide some added value (utility) and does this by engaging in behaviour over time, during which data is exchanged with other systems over communication channels. (We restrict our attention to systems that exchange data.) And these services are delivered at a certain level of quality.

The GRAAL framework contains three other dimensions. The decomposition dimension says that each system is decomposed into subsystems, that are encapsulated in it. The system life dimension says that each system goes through stages in its life, from conception to disposal. The refinement dimension says that each system can be described at different levels of abstraction, from very abstract (few details) to very detailed.

3.3.1 THE ZACHMAN FRAMEWORK [13,20].

The rows in the Zachman framework represent the perspectives of different roles on the system (Fig. 1(a)). The planner considers the scope of the system in relation to the environment, the owner considers the role of the system in the enterprise, the designer considers the software needed to achieve the business goals, and the builder considers the infrastructure needed to build the system. So far, this corresponds to layers in the GRAAL framework. The subcontractor role moves however to subsystems of a system, and this is moving along another GRAAL dimension, namely the decomposition dimension. Because decomposition can be done at any level in the GRAAL framework, it should not be placed at the lowest level only, as Zachman does.

The data, function and network aspects of Zachman correspond to the GRAAL aspects of data, service and communication. Zachman's time aspect corresponds roughly to the behaviour aspect in GRAAL, because the behaviour as a func- tional property of a system is the ordering of product interactions in time [21, p. 40]. The people aspect is represented in GRAAL's enterprise layer, because people are part of the enterprise and therefore of the business aggregation hierarchy. Finally, the motivation aspect from Zachman corresponds to the utility aspect of GRAAL, because the utility of an entity at any layer for entities at higher layers is the motivation why this lower-level entity exists. We conclude that the Zachman framework can be mapped into the GRAAL framework.

3.3.2 THE FOUR-DOMAIN ARCHITECTURE [22]

This framework distinguishes four domains (Fig. 1(b)). The process domain includes processes, procedures, business tools and dependencies required to support business functions. This corresponds to the behaviour aspect of entities at the enterprise level. The information/knowledge domain includes business rules, data, all types of information, definitions, interrelationships, etc. and corresponds to the aspects communication and data, in the GRAAL framework, also at the enterprise level. The infrastructure domain includes facilities, hardware, system software, networks, etc. This corresponds to the software infrastructure and physical infrastructure layers from GRAAL. It spans even the quality aspect of GRAAL, because reliability and availability are elements expected to be documented within the infrastructure domain. The organisation domain includes business people and their roles and responsibilities, organisational structure as well as interrelationships to all kind of stakeholders. This corresponds partly to the utility and service aspect of enterprise-layer entities in the GRAAL framework (who does what for whom?). The organisational structure and relationships are part of the system decomposition dimension of GRAAL, which is not shown in our figures. Note that the Four-Domain Architecture was developed for managing EAs and to support frameworks such as the Zachman's. Iyer and Gottlieb also distinguish architecture design from architecture use and split each of the cells of Zachman's framework into two subcells, corresponding to these two stages in the life of an architecture. This corresponds to the system life dimension of GRAAL.

3.3.3 TOGAF [14]

TOGAF is, compared to all other frameworks presented in this paper, the most comprehensive one, because TOGAF offers a complete guide for the development of an EA and comes up with an architectural development method. Such a step-by-step guide is absent at Zachman and GRAAL. TOGAF distinguishes four kinds of architectures, namely business architecture, data architecture, application architecture and technology architecture, where data architecture and application architecture is sometimes referred to as information systems architecture. We consider these architectures as the highest-level building blocks to be documented. The mapping to GRAAL is straightforward (Fig. 1(c)).

3.3.4 RM-ODP [23]

RM-ODP (Reference model for open distributed processing) provides five viewpoints (Fig. 1(d)), where a viewpoint is a subdivision of the specification of a complete system, established to bring together those particu- lar pieces of information relevant to some particular area of concern during the design of the system" [24]. A viewpoint in RM-ODP can span aspects of one or more viewpoints in GRAAL and vice versa. The RM-ODP enterprise viewpoint focuses on the purpose, scope and policies of the system and provides the overall environment in which the system will be built. This corresponds roughly to the enterprise and enterprise environment layers in the GRAAL framework. However, the RM-ODP enterprise viewpoint can also specify more technical entities like operating systems or database systems [25], which lie in GRAAL on the software infrastructure layer. This is not represented in Fig. 1(d).

The information viewpoint is a viewpoint on the system and its environment that focuses on the semantics of the information and information processing performed. This corresponds roughly to the data aspect on the enterprise system layer in GRAAL. The computational viewpoint enables distribution through functional decomposition of the system into objects which interact at interfaces. In GRAAL this viewpoint corresponds to the decomposition dimension. It is not shown in Fig. 1(d). The engineering viewpoint focuses on the mechanisms and functions required to support distributed interaction between objects in the system, which corresponds roughly to the communication aspect on the same layer. The technology viewpoint focuses on the choice of technology in the system. It is used to

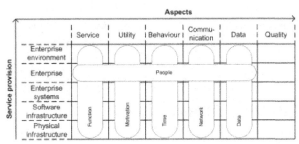

(a) Comparison between GRAAL and the Zachman framework

(b) Comparison between GRAAL and the Four-Domain Architecture

(c) Comparison between GRAAL and TOGAF

(d) Comparison between GRAAL and RM-ODP

FIGURE 1: Comparison between GRAAL and other well know frameworks

build technology specifications of particular configurations of hardware elements, software elements and networks. The technology viewpoint corresponds to the physical infrastructure and software infrastructure layers in GRAAL.

3.4 DISCUSSION AND FURTHER WORK

The paper described briefly five frameworks and compared them. The basis of the comparison was the GRAAL framework. This section summarizes our results.

3.4.1 ABSTRACTION MECHANISMS

The frameworks use different abstraction mechanisms. GRAAL and Zachman use dimensions, but the other frameworks populate some of these dimensions. The GRAAL dimensions Aspects and Service layers correspond roughly to the Zachman dimensions Aspects and Perspectives. The GRAAL dimensions Decomposition, System life, and Refinement are not mentioned as such by Zachman, although he does include a decomposition perspective (subcontractor). As we have seen, the abstraction mechanisms of the other frameworks, namely the domains, architectures and viewpoints are mostly a combination of the system aspects dimension and the service layers in GRAAL. The abstraction mechanisms used by GRAAL and Zachman are at a metalevel with respect to the others, which explains the relationship between the different frameworks. EAFs at a higher level of abstraction can integrate frameworks that use lower level abstraction mechanisms.

3.4.2 THE INTEGRATED ENTERPRISE ARCHITECTURE FRAMEWORK

Our analysis makes clear that to find an integrated framework, we can build on GRAAL. We just extend GRAAL with a number of domains, which we place on the appropriate layers. Figure 2 shows the resulting integrated EAF (IEAF). In the IEAF we merged the two infrastructure layers for simplicity. Many EAFs, like RM ODP or the Four-Domain architecture also do not have a distinction between the two on their highest level of abstraction. This serves to integrate such frameworks more easily. The enterprise network domain contains sets of interacting business actors that are profit-and-loss responsible, such as independent businesses, or business units within a large corporation. Therefore it is especially interesting for networked business constellations. The organisation structure domain of the IEAF contains the decomposition of an organisation into whatever elements are recognised, such as units, departments, employee roles, etc. The business process domain consists of business processes and the communication and information domains consists of human or automated communication channels and the information passed through them. The services domain consists of IT services, and the behaviour domain consists of software behaviour. The infrastructure domain consists of all software and hardware needed to facilitate the higher layers. Note that the IEAF has the advantage that each layer is not fix but °exible. This means that, if necessary, additional domains can be added to an IEAF layer, which goes so far that domains can even span several layers as shown in Fig. 2. Because the layers are not fix they can be viewed as dimensions.

The implication of the IEAF for business-IT alignment is that each domain is an area of design knowledge, and that the alignment problem must be decom- posed into these domains. In the case of interorganisational business-IT alignment, the additional implication is that the EAFs of the partner companies can be mapped to each other using the IEAF as common reference. Our research implies that each of these EAFs can be mapped to the IEAF and that this is the way the EAFs should be mapped to each other. Further, the IEAF forms a basis for communication between different businesses cooperating in a networked value constellation.

3.4.3 EVALUATION

In design science the evaluation phase is very important. "Design science addresses research through the building and evaluation of artifacts" [26].

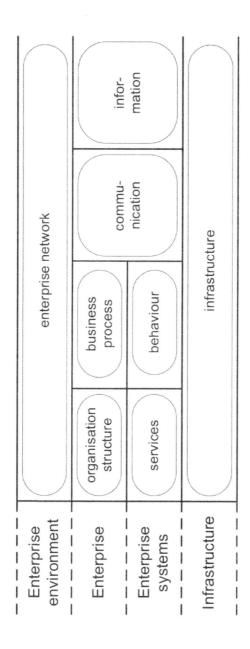

FIGURE 2: An Integrated Enterprise Architecture Framework (IEAF) for networked business constellations

The GRAAL framework, as our built artifact, was used to analyse the EAFs of five different organisations in the Netherlands, each having a different EAF. All the information from those EAFs could be incorporated without information loss into the GRAAL framework. We consider this as a first validation of the GRAAL framework. The IEAF, as our simplified artifact, inherited most of its documentation structure from the GRAAL framework. For the evaluation of our artifacts case studies were used, like described by Hevner et al. [26, page 86].

As we have seen during the comparisons in Sec. 3, the abstraction mechanisms, namely domains, architectures and viewpoints of the other frameworks are mostly a combination of the systems aspect dimension and the service layers in GRAAL. Therefore dimensions, like the layers in the IEAF, are at a higher level of abstraction than the abstraction mechanisms used by most of the other EAFs. We also identified that the frameworks use a fixed number of abstraction mechanisms. The number of domains in the IEAF (which is not fixed) was identified and situated on the appropriate layers after having examined the frameworks used by our industrial partners for documenting their enterprise architectures. They covered all the information documented. Therefore it is true that it is an integrated EAF.

Corroboration of the IEAF is also provided by an independent investigation of IT architect roles [27]. The architect roles distinguished by most companies coincide with the domains identified in the IEAF. Further validation is provided by a preliminary investigation of the EAFs of several companies that were compared with the IEAF and could all be mapped to it without loss of information.

3.4.4 SUMMARY AND FUTURE RESEARCH

The frameworks described in this paper use different abstraction mechanisms, but can nevertheless be mapped into the GRAAL framework with some degree of approximation. The result is an integrated EAF (IEAF). The IEAF serves as a reference point between different organisations, and enables them to understand eachother's frameworks. The presence of the domains provide an additional useful utility. In the future we will investigate cross-organisational integration problems using the IEAF as our conceptual framework.

REFERENCES

1. Langenberg, K., Wegmann, A.: Enterprise Architecture: What Aspects is Current Research Targeting? Technical report, EPFL (2004) http://icwww.epfl.ch/publications/documents/IC_TECH_REPORT_200477.pdf.
2. Schekkerman, J.: How to survive in the jungle of Enterprise Architecture Frameworks. Trafford (2004)
3. A. Tang and J. Han and P. Chen: A Comparative Analysis of Architecture Frameworks. Technical report, Swinburne University of Technology (2004) http://www.swin.edu.au/ict/research/technicalreports/2004/SUTIT-TR2004.01.pdf.
4. Blanchard, B.S., Fabrycky, W.J.: Systems Engineering and Analysis. Prentice-Hall (1990)
5. IEEE Architecture Working Group: Definition of Architecture. http://www.pithecanthropus.com/~awg/public_html/ieee-1471-faq.html (2002)
6. Bass, L., Clements, P., Kazman, R.: Software Architecture in Practice. Addison-Wesley (1998) Integrated Enterprise Architecture Framework 9
7. Wieringa, R.: Architecture is Structure plus Synergy. http://graal.ewi.utwente.nl/WhitePapers/Architecture/architecture.htm (2004)
8. West, D., Bittner, K., Glenn, E.: Ingredients for Building Effective Enterprise Architectures. http://www-128.ibm.com/developerworks/rational/library/content/RationalEdge/nov02/EnterpriseArchitectures_TheRationalEdge_Nov2002.pdf (2002)
9. Schekkerman, J.: The Economic Benefits of Enterprise Architecture. Trafford (2005)
10. Perks, C., Beveridge, T.: Guide to Enterprise IT Architecture. Springer (2003)
11. Bernard, S.A.: An Introduction to Enterprise Architecture. Authorhouse, Bloomington, Indiana (2004)
12. Wieringa, R.: The GRAAL Architecture Framework. http://graal.ewi.utwente.nl/WhitePapers/Framework/framework.htm (2004)
13. Zachman, J.: A framework for information systems architecture. IBM Systems Journal 26(3) (1987)
14. The Open Group: The Open Group Architecture Framework (TOGAF) - Version 8, Enterprise Edition (2002)
15. van Eck, P., Blanken, H., Fokkinga, M., Grefen, P., Wieringa, R.: A Conceptual Framework for Architecture Alignment Guidelines. http://graal.ewi.utwente.nl/GRAAL_whitepaper_20021017.pdf (2002)
16. Wieringa, R., Blanken, H., Fokkinga, M., Grefen, P.: Aligning Application Architecture to the Business Context. In: Conference on Advanced Information System Engineering (CAiSE 03), Springer (2003) 209{225 LNCS 2681.
17. Wieringa, R., van Eck, P., Krukkert, D.: Architecture Alignment. In Lankhorst, M., ed.: Enterprise Architecture at Work. Springer, Berlin (2005)
18. Wieringa, R.: Requirements Engineering. John Wiley, Chichester, England (1996)

19. Wieringa, R.: A Survey of Structured and Object-Oriented Software Specification Methods and Techniques. ACM Computing Surveys 30(4) (1998) 459{527

20. Sowa, J., Zachman, J.: Extending and formalizing the framework for information systems architecture. IBM Systems Journal 31(3) (1992) 590{616

21. Wieringa, R.: Reactive Systems. Morgan Kaufmann, San Francisco (2003)

22. Iyer, B., Gottlieb, R.: The Four-Domain Architecture: An approach to support enterprise architecture design. IBM Systems Journal 43(3) (2004) 587{597

23. International Organization for Standardization: RM ODP - Open Distributed Processing Reference Model - ISO/IEC 10746-1, ISO/IEC 10746-2, ISO/IEC 10746-3 and ISO/IEC 10746-4. (http://isotc.iso.org/livelink/livelink/fetch/2000/2489/Ittf_Home/PubliclyAvailableStandards.htm)

24. Vallecillo, A.: RM-ODP: The ISO Reference Model for Open Distributed Processing. (citeseer.ist.psu.edu/277001.html)

25. Joyner, I.: Open distributed processing: Unplugged! http://homepages.tig.com.au/~ijoyner/ODPUnplugged.html (1997)

26. Hevner, A., March, S., Park, J.: Design Science in Information Systems Research. MIS Quarterly 28(1) (2004) 75{105

27. Wieringa, R.: An inventory of architect roles and competencies. In: IT Practitioners Conference, Barcelona (2006) http://www.opengroup.org/public/member/proceedings/q106/.

CHAPTER 4

ENTERPRISE ARCHITECTURE: NEW BUSINESS VALUE PERSPECTIVES

M. DE VRIES and A. C. J. VAN RENSBURG

4.1 INTRODUCTION

In the 1970s and 1980s, business processes were redesigned roughly once every seven years. This provided ample time to alter information systems. In the 1990s the rate of change started to increase rapidly, and information systems lagged behind. Today IT departments struggle to keep up with the rapid change of business processes (Wagter, van den Berg, Luijpers & van Steenberg [1]).

Looking at the history of enterprise architecture, different eras become apparent:

The mainframe era required a centralised approach. The need for EA was minimal as a limited set of resources had to be managed. However, the centralised IT departments failed to meet the demands of business users.

As technology evolved and became more accessible, organisational units began to evolve, each one deploying its own systems to improve service. The decentralised approach led to an expansion of system complexity and a loss of functionality between departments.

This chapter was originally published under the Creative Commons License or equivalent. de Vries M and van Rensburg ACJ. Enterprise Architecture: New Business Value Perspectives. South African Journal of Industrial Engineering *19,1 (2008).*

The loss of control led to ERP (Enterprise Resource Planning) systems as a solution to complex, multiple, fragmented, and non-interoperable legacy systems. But these ERP systems were very costly, demanded ongoing maintenance, and required major system modifications to address business processes. Also, mergers and acquisitions once again diversified and complicated the system landscape (Theuerkorn [2]).

The complex system landscapes of today led to the need to manage the evolution of system and technology environments, which in turn led to the emergence of a new profession called Enterprise Architecture (EA).

4.2 ENTERPRISE ARCHITECTURE

4.2.1 DEFINITION AND VALUE PROPOSITION

EA is a management practice that aims at improving performance of enterprises. EA gained impetus with the USA Clinger-Cohen Act (CCA) of 1996. This act assigned the CIO the responsibility of "developing, maintaining, and facilitating the implementation of a sound and integrated Information Technology Architecture" (Schekkerman [3]). The Information Technology Architecture (ITA) had to ensure that existing information technology was maintained and new information technology was acquired to achieve the agency's strategic goals and information resources management goals.

In the past, EA was the responsibility of the IT unit(s) in a company. However, many IT architecture efforts were remote from reality, and were represented in overlycomplex diagrams. Companies defined strategy piece-meal, which delivered separate IT solutions for each strategic initiative, rather than delivering IT capabilities. This resulted in IT being a constant bottleneck. Standalone systems were created, causing poor customer, supplier, and employee process coordination. Data were also patchy, error-prone and not up to date (Ross, Weill, & Robertson [4]). Currently companies realise that EA is not an IT issue, but a business issue.

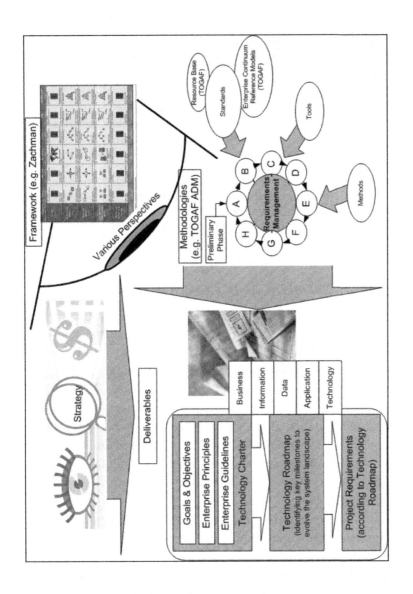

FIGURE 1: Conceptual framework for EA

According to Ross et al [4] EA should strive at providing "...the high-level logic for business processes and IT capabilities". They realised that EA is not so much to achieve a particular end state as it is to serve as a blueprint for a company's direction. EA should provide a "holistic and integrated view of the strategic direction, business practices, information flows, and technology resources" of the company (Bernard [5]). The main EA concepts are:

- strategic planning deliverables that direct EA objectives;
- a technology charter (enterprise objectives, principles and guidelines and derived technology objectives, principles and guidelines);
- a technology roadmap (milestones to evolve the system landscape) that is derived from the technology charter;
- project requirements that are identified to meet the technology milestones;
- a methodology (including methods, tools and standards) that is used to support the evolution of the system landscape; and
- a framework that classifies EA models to communicate to various stakeholder groups.

4.2.2 RESEARCH PROBLEM AND GOAL OF THE STUDY

Although EA offers numerous benefits and value propositions, many organisations perceived EA as another black hole that provided a low return on investment if measured by the traditional financial measurement system. EA practitioners tried to improve the EA practices, frameworks, methodologies, and tools to reduce the number of artefacts and/or to accelerate the implementation of EA governance mechanisms (Theuerkorn [2] and Wagter et al [1]). Although Ross et al [4] realised the importance of using EA as a blueprint in directing a company, they still failed to demonstrate how EA objectives should be measured or how they could be converted to tangible value.

EA creates value on different management levels in the organisation across multiple domains (e.g. governance, strategy, business processes, information, applications, technology, workforce management, security, and standards). Literature also indicates that EA as a management programme complements various other areas. Some of these include strategic planning, strategy execution, quality management, IT governance (e.g.

complementing COBIT – Control Objectives for Information and related Technology), IT Service Delivery and Support (e.g. supporting the key processes of ITIL – IT Infrastructure Library), and IT Implementation (supporting the implementation of best practices) (Lankhorst [6]).

An integrated approach is required to demonstrate how EA creates tangible value on both enterprise level and strategic business unit (SBU) level, across different domains. Kaplan & Norton [7,8,9] provide numerous tools to create synergies, alignment, and integration of intangible assets on different organisational management levels. The next section describes how intangible assets create value on both an enterprise level and SBU level. The management level perspectives are then used to discuss generic EA objectives, their relation to other strategic objectives, and links to performance measurement.

4.3 DIFFERENT PERSPECTIVES ON VALUE CREATION

During the same time that EA was initiated as a management practice, Kaplan & Norton [7] identified a major shortcoming in the traditional measurement systems. These overemphasised achieving and maintaining short-term financial results (i.e. high ROI) that lead to overinvestment in short-term fixes and underinvestment in long-term value creation – especially underinvestment in intangible and intellectual assets that generate future growth. Kaplan & Norton [7] contributed to the new understanding of what creates value for organisations. Intangible assets (e.g. motivated/ skilled employees, responsive and predictable internal processes, and satisfied customers) are some of the most important sources of long-term value creation. Unfortunately traditional financial measurement systems merely focused on tangible assets.

A more balanced measurement system was proposed to incorporate four perspectives: financial, customer, internal processes, and learning and growth (Kaplan & Norton [7]). They believed that companies had to invest in these four domains to create both short-term financial improvement and long-term profitable growth. They also realised that companies that wanted to survive and prosper in the information age competition had to use measurement and management systems that are derived from their strategies and capabilities. The balanced scorecard (BSC) was developed to account for the different measurement perspectives, and provided a systematic process of implementing and obtaining feedback about strategy.

With the balanced perspective on organisational measurement, Kaplan & Norton [8] posed new value-creation perceptions concerning intangible assets. They assert that intangible assets are usually bundled, seldom create value by themselves, and do not have a value that can be isolated from the organisational context and strategy. Intangible assets are expected to help the organisation accomplish the strategy; hence action plans should be aligned around strategic themes. Integrated bundles of investments should be linked to the strategic themes instead of managing standalone projects. Each investment or initiative is "only an ingredient in the bigger recipe" (Kaplan & Norton [8]). Economic justification should only be determined by evaluating the return from the entire portfolio of investments in intangible assets. Kaplan & Norton [9] also realised that the conglomerate and multidivisional organisation structures of today do not only achieve growth through expansions from the core business, technologies, and capabilities, but also through acquiring and merging unrelated businesses. Senior executives of these conglomerates need to add superior knowledge and skills to the newly-owned organisations to make the merger or acquisition worthwhile. The value of the collection of companies should thus be more than if the companies operated independently without the benefit of the corporate office.

Value creation on a corporate level consequently differs from value creation at a strategic business unit (SBU) level. Corporate offices need to create enterprise-derived value, creating alignment and synergy between SBUs. On the other hand, SBUs need to show how their internal capabilities and assets are used to create customer-derived value. Different perspectives on value creation are demonstrated in Figure 2. The next two sections emphasise value-creation strategies at a corporate level and related EA value propositions.

4.4 STRATEGIES TO CREATE CORPORATE SYNERGIES

Various strategies could be used to create synergies on a corporate level. A few examples include:

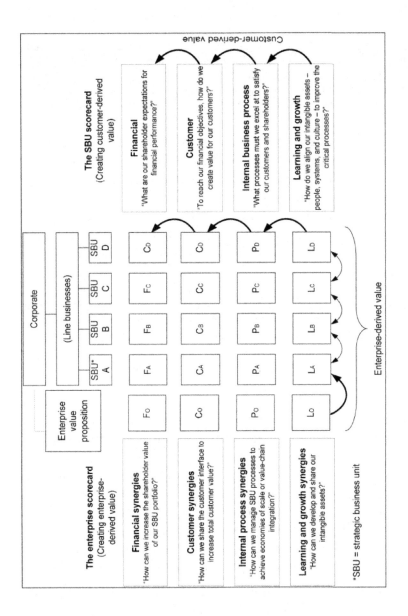

FIGURE 2: Building the enterprise scorecard (Kaplan & Norton [9])

- Financial perspective: using effective merger and acquisition policies or generating synergies by using centralised resource allocation and financial management.
- Customer perspective: leveraging common brand or customer relationships across different business units and retail channels.
- Internal process perspective: gaining economies of scale by sharing common processes and services or gaining economies of scope by integrating business units across an industry value chain.
- Learning and growth: creating synergies by sharing human, information, and organisation capital across multiple units (Kaplan & Norton [9]).

Ross et al [4] created an EA value proposition primarily in terms of the internal process perspective (gaining economies of scale by sharing common processes/ services or integrating processes). They believe that EA objectives should be defined with regards to the enterprise operating model. The operating model is an actionable view of the company's strategy that outlines the expectations for integration and standardisation across business units. The operating model is thus defined in terms of two dimensions:

- Business process standardisation – the extent to which the company benefits by having business units run their operations the same way.
- Business process integration – the extent to which different business units are dependent on one another for accurate and timely data. Different combinations of these dimensions are portrayed in Figure 3.

The researcher used the operating model parameters to identify generic EA objectives within the areas of process management, data sharing, application management, and infrastructure. Figure 4 provides a proposed set of EA objectives for a required operating state.

Ross et al [4] also complemented this model with a methodology for implementing the required operating model. The methodology acknowledges different levels of EA maturity and suggests a phased (project-by-project) approach. The methodology also includes some element of integration with other intangible assets (such as organisational learning and leadership requirements). It is proposed that the operating model approach could be further enhanced by mapping the EA objectives (related to a specific operating model) to a corporate strategy map.

FIGURE 3: Characteristics of four operating models (Ross et al [4])

To demonstrate the concept, the author mapped EA objectives for the unification operating model to a corporate strategy map. Additional EA objectives (as found in Schekkerman [3], Theuerkorn [2], Ross et al [4], Wagter et al [1], Boar [10], Iyamu [11]) were also included to demonstrate a comprehensive strategy map for EA capital (see Figure 5). The strategy map indicates that most EA objectives support the creation of process synergies and learning & growth synergies between SBUs.

Architecture Dimension	EA Objective	Operating Model			
		Diversification	Coordination	Replication	Unification
Process Management	To centralise processes (versus business unit autonomy)	No	No	Yes	Yes
	To centralise standard process designs (versus control by separate business units)	No	No	Yes	Yes
	To integrate processes between different business units.	No	Yes	No	Yes
Data Sharing	To share customer data	No		No	Yes
	To share product data	Indifferent	Either	Indifferent	Indifferent
	To share supplier data	No		Indifferent	Yes
Application Management	To centralise IT application decision making (i.e. not prescribed by separate business units)	No	No	Yes	Yes
Infra-structure	To create consensus processes for designing IT infrastructure services	No	Yes	Yes	Yes

FIGURE 4: EA objectives per operating model (based on the four operating models identified by Ross et al [4])

4.5 USING STRATEGIC THEMES TO CREATE CORPORATE SYNERGIES

Kaplan & Norton [9] also report the effective use of corporate strategic themes to create synergy between business units. Examples of strategic themes include operational excellence, and complete solutions to targeted customers. The corporate strategic themes are used in combination with the corporate BSC to cascade strategic objectives to individual SBUs. SBU managers are then obliged to phase out local projects that are not contributing to one or more of the strategic themes.

The purpose is to create alignment and integration among the diverse and dispersed business units. Strategic themes in the corporate scorecard have the ability of allowing decentralised units to seek local gains while still contributing to corporatewide objectives. Due to the diversity of the business units, not all units are expected to contribute to all the themes. Kaplan & Norton [8]) agree with Treacy & Wiersma [12] that a single SBU usually focuses on one strategic theme, as it is impossible to excel in all areas simultaneously.

They identified three main strategic areas:

1. Operational excellence: the best total costs.
2. Customer intimacy: the best total solution/customised mix of products and services to solve customers' problems.
3. Product leadership: excel in the offering of products and services.

Hax & Wilde [13] articulated a fourth strategic area, called 'system lock-in', in which companies provide a system platform that becomes an industry standard.

Kaplan & Norton [8] combined the strategic areas with the four scorecard perspectives (financial, customer, internal processes, and learning & growth) to discuss the different objectives that would be required for each strategic focus area. An SBU usually selects one strategic focus area (e.g. operational excellence) to direct the selection of primary strategic objectives within the four scorecard perspectives. Finally, strategy maps are used to link the different objectives (within the four perspectives) in cause-and-effect relationships. The learning and growth perspective objectives (cause) need to support the internal process objectives (effect). Furthermore, the internal process objectives (cause) need to enable the achievement of the customer objectives (effect), while the customer objectives (cause) need to contribute to the accomplishment of the financial objectives (effect).

Kaplan & Norton [8] identified four categories of value-creating internal processes:

- Operations management processes;
- Customer management processes;
- Innovation processes; and
- Regulatory and social processes.

Different process objectives are required for each strategic focus area. The different process objectives are discussed and argued in the next two sections.

4.5.1 PROCESS OBJECTIVES PER FOCUS AREA

4.5.1.1 OPERATIONAL EXCELLENCE/LOW TOTAL COST

This focus area requires highly competitive processes combined with consistent quality, ease and speed of purchase, and excellent product selection. Customer management processes require ease of access for customers. Accessible order processes and superb post-sale services are required. The SBU would perform market research to understand the most preferred range of products and services by the largest segments of customers. SBUs that pursue this strategic focus are product followers, not leaders, and do not invest a great deal in product and service innovation. They innovate on processes rather than products. They also emphasise regulatory and social processes to avoid accidents and environmental incidents that are costly to the company (Norton & Kaplan [8]).

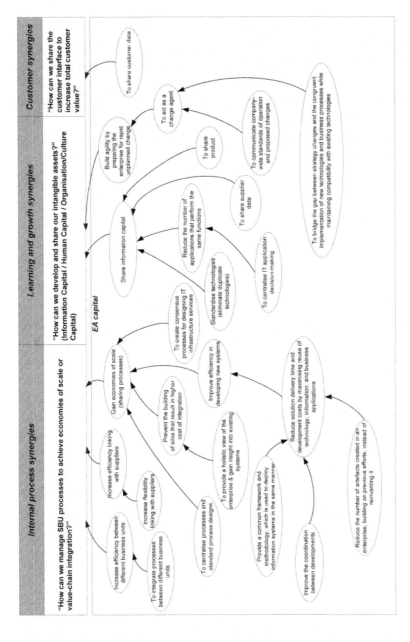

FIGURE 5: EA objectives linked to a corporate model

4.5.1.2 CUSTOMER INTIMACY

SBUs offer 'customer solutions' value propositions that stress objectives related to the completeness of the solution (selling multiple bundled products and services), exceptional service, and a quality relationship. The company tries to deepen the relationship with current customers to sell multiple related products and services. Innovation processes focus on finding new ways to create value for customers, such as providing alternative ways for customers to access the company's products and services. Regulatory and social processes are focused on gaining regulatory approval to offer services that cut across traditional industry barriers (Norton & Kaplan [8]).

4.5.1.3 PRODUCT LEADERSHIP

SBUs that emphasise product leadership provide products with features and functionality that leading-edge customers are willing to pay for. The value proposition includes outstanding performance, accuracy, size, or power consumption. Product leadership SBUs extend superior functionality of products into multiple market segments. Their key internal processes are in the innovation cluster. Flexibility and improvement of operating processes are more important than low-cost production. Customer management objectives include the identification of requirements from leading-edge customers, as well as educating customers about the benefits of new/advanced products. Regulatory and social processes must avoid the adverse side effects that may occur when they introduce new products. SBUs should emphasise objectives related to product safety, employee and customer health, and environmental impacts of new products (Norton & Kaplan [8]).

4.5.1.4 LOCK-IN

The new economy information-based industries, such as computer hardware, software, internet, and telecommunications, led to lock-in strategies, creating high switching costs for customers. SBUs that pursue a lock-in strategy require powerful innovation processes. They need to develop a proprietary product or protected standard that serves as the basis for lock-in. As complementors provide a source of sustainability, SBUs need to acquire and retain complementors. They need to attract new customers by lowering their switching costs. High margins from successful lockin remove the pressure of having the most efficient operating processes. Two critical regulatory objectives must be pursued: protecting proprietary products from imitation and use by competitors, and preventing product-use by unauthorised customers (Norton & Kaplan [8]).

4.5.2 OPERATIONAL EXCELLENCE: THE DOMINANT THEME?

In the past, many SBUs (as well as industrial engineers) focused primarily on the area of operational excellence. They emphasised the improvement of existing processes that would lead to short term financial improvement. Various process/quality/productivity improvement programmes were developed and implemented (e.g. ISO 9001, Six Sigma, European Foundation for Quality Management, Capability Maturity Models, and Supply Chain Management programs). Norton & Kaplan [7] believe, however, that some SBUs may benefit more in the long term by anticipating customer needs or delivering new services that target customers would value. Conversely, other business units may benefit most if they focus on their innovation processes, creating entirely new products and services.

Ross et al [4] believe that a SBU needs to recognise its core operations and digitise these to enable organisations to exploit their foundation for execution, which should lead to agility and profitable growth. The rationale is that digitising core business processes helps in automating some routine activities, making sure that these are done reliably and predictably. Management could then focus their attention on higher-order processes (serving customers, developing new products, seizing new opportunities). Their definition of core business processes thus implies a focus on a single process area, namely operational management. Norton & Kaplan [8] agree that many SBUs first stabilise their operations and delivery processes to produce consistent output in conformity to specifications. The definition of quality then shifts from conforming to specifications to meeting customers' expectations. One could reason that an organisation's strategic focus area will change to customer intimacy/product leadership/lock-in as the organisation reaches a certain level of maturity.

4.6 A NEW VALUE-CREATION APPROACH

Intangible assets do not have a value that can be isolated from the organisational context and strategy. They are expected to help the organisation to accomplish the strategy. EA, used in combination with the BSC, creates the context on a corporate level to ensure that intangible assets (especially value-creating processes and information capital) are integrated with other intangible assets and aligned around the strategic themes of the organisation. EA initiatives should be treated as part of an integrated bundle of investments, linked to strategic themes, instead of managing them as standalone projects. Economic justification should only be determined by evaluating the return from the entire portfolio of investment in intangible assets.

EA thus has the potential to produce corporate synergies between strategic business units, especially creating internal process synergies and learning & growth synergies. Furthermore, EA objectives primarily support the operational excellence theme, which may be further differentiated using operating model parameters. EA objectives on the corporate level could then be cascaded to SBU level and division level to ensure alignment with the corporate scorecard objectives (Figure 6).

4.7 AN ANGLO PLATINUM EXAMPLE

Anglo Platinum is the world's leading primary producer of platinum, and accounts for about 38% of the world's newly-mined production. Operations comprise seven mines, three smelters, a base metals refinery, and a precious metals refinery, situated in the Bushveld Complex north-west and north-east of Johannesburg. Anglo Platinum embarked on one of the most successful EA initiatives in the world; some critics consider them to be in the top 5% globally [14]. The following narrative explains how Anglo Platinum followed an approach similar to the proposed valuecreation approach.

During 2004 corporate management restructured and effectively decided to move away from a diversified operating model to a replication operating model by centralising control over business process designs and providing clear corporate guidelines. Moving towards a replication operating model would enhance process synergies between the different business units. Corporate management also defined strategic themes to drive their strategic objectives. The most prominent themes were operational excellence and social upliftment.

The Group Information Collaboration Technology (ICT) division of Anglo Platinum received the mandate to identify value-creation process objectives as well as information and technology objectives according to the required operating model and aligned to the operational excellence theme. Contrary to the traditional IT department role of service provider, the role of Group ICT was to act as a decisionmaking body in defining business process requirements and information quality/requirements. The division, employing 120 people—predominantly business managers and a few technology experts—embarked on strategic work sessions to define strategic drivers and themes in directing their divisional strategic objectives. Finally, a strategy map was used to demonstrate the cause-and-effect links between their strategic objectives. Although these objectives embodied multiple EA objectives, EA did not feature as a separate theme or management approach. The prominent EA objectives were:

• Standardisation of business processes to reduce complexity.
• Governance of standardised processes to ensure implementation and adherence on an operational level.

The EA objectives primarily focused on the business architecture layer of the enterprise, and were combined with other projects (e.g. business process improvement and information value-chain initiatives) to form an integrated bundle of projects aligned to the operational excellence strategic theme. The standardisation and stabilisation of business processes provided a platform for continuous improvement. Simulation is currently considered as an innovation mechanism to drive continuous improvement.

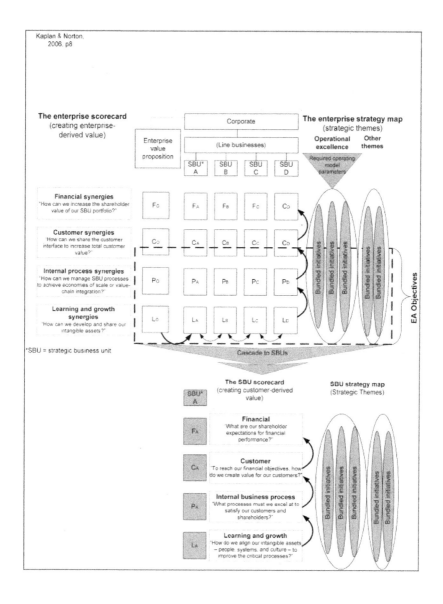

FIGURE 6: A value-creation approach for EA

4.8 CONCLUSION

In the past, many organisations overemphasised short-term return on investment rather than long-term financial improvement. This also had a negative effect on the perceived value of EA. Being an intangible asset, EA has the potential to unlock value if perceived in the context of strategy and long-term profitable growth. The corporate BSC and strategy map were used to demonstrate EA potential in creating process synergies and learning & growth synergies. It was also found that EA objectives primarily support an operational excellence theme, while the subset of EA objectives is mainly determined by the operating model of the specific enterprise.

This research provided the context for EA planning in an enterprise. The suggested approach requires alignment between different intangible assets according to the theme of operational excellence. Further research will be done to demonstrate the alignment of EA with other intangible assets as part of the planning stages of theme-related initiatives.

REFERENCES

1. Wagter, R., van den Berg, M., Luijpers, J. & van Steenberg, M. 2005. Dynamic Enterprise Architecture – How to make it work. John Wiley & Sons, Inc., Hoboken, New Jersey.
2. Theuerkorn, F. 2005. Lightweight Enterprise Architectures. Auerbach Publications, New York.
3. Schekkerman, J. 2004. How to survive in the jungle of Enterprise Architecture frameworks. Second edition, Trafford Publishing, Canada.
4. Ross, J.W., Weill, P. & Robertson, D.C. 2006. Enterprise Architecture as strategy: Creating a foundation for business execution. Harvard Business School Press, Boston, Massachusetts.
5. Bernard, S.A. 2005. An introduction to Enterprise Architecture EA3. Second edition, Authorhouse, Bloomington, USA.
6. Lankhorst, M. 2005. Enterprise Architecture at work. Springer-Verlag, Berlin, Germany.
7. Kaplan, R.S. & Norton, D.P. 1996. The Balanced Scorecard. Harvard Business School Press, Boston, Massachusetts.
8. Kaplan, R.S. & Norton, D.P. 2004. Strategy maps: Converting intangible assets into tangible outcomes. Harvard Business School Press, Boston, Massachusetts.

9. Kaplan, R.S. & Norton, D.P. 2006. Alignment – using the Balanced Scorecard to create corporate strategies. Harvard Business School Press, Boston, Massachusetts. 2006.
10. Boar, B. H. 1999. Constructing blueprints for Enterprise IT Architectures. John Wiley, New York.
11. Iyamu, T. 2002. Enterprise Architecture as an enterprise change agent. Thesis in partial fulfilment of the requirements for Masters in Information Management, University of the Western Cape.
12. Treacy M. & Wiersma, F. 1995. The discipline of market leaders: Choose your customers, narrow your focus, dominate your market. HarperCollins, London.
13. Hax, A. & Wilde, D. 2001. The Delta Project: Discovering new sources of profitability in a networked economy. New York, Palgrave.
14. Macgregor, S. 2007. EA enduring competitive advantage. Newsletter. June.
15. Scott, A.B. 2005. An introduction to Enterprise Architecture, second edition. Authorhouse, Bloomington, United States of America.

PART II

OPEN ISSUES AND NOVEL IDEAS

CHAPTER 5

PATTERN-ORIENTED APPROACH FOR ENTERPRISE ARCHITECTURE: TOGAF FRAMEWORK

MOHAMED TALEB and OMAR CHERKAOUI

5.1 INTRODUCTION

In recent years, many industrial firms have adopted architectures called enterprise architecture (EA). The Enterprise Architecture has matured from offering a lot of functionalities to like providing a clear representation of business processes and information systems, improving the IT governance, planning changes and optimizing resources.

Several definitions have been suggested by several authors. For example, The Institute for Enterprise Architecture Developments [1] "Enterprise Architecture is a complete expression of the enterprise; a master plan which acts as a collaboration force between aspects of business planning such as goals, visions, strategies and governance principles; aspects of business operations such as business terms, organization structures, processes and data; aspects of automation such as information systems and databases; and the enabling technological infrastructure of the business such as computers, operating systems and networks," Giachetti and MIT Center [2,3] "Enterprise Architecture is a rigorous description of the structure of

This chapter was originally published under the Creative Commons License or equivalent. Taleb M and Cherkaoui O. Pattern-Oriented Approach for Enterprise Architecture: TOGAF Framework. Journal of Software Engineering and Applications, 5,1 (2012), pp. 45-50. DOI: 10.4236/jsea.2012.51008.

an enterprise, which comprises enterprise components (business entities), the externally visible properties of those components, and the relationships (i.e. the behavior) between them. Enterprise Architecture describes the terminology, the composition of enterprise components, and their relationships with the external environment, and the guiding principles for the requirement (analysis), design, and evolution of an enterprise", the Enterprise Architecture Center of Excellence [4] "Enterprise Architecture explicitly describing an organization through a set of independent, non-redundant artifacts, defining how these artifacts interrelate with each other, and developing a set of prioritized, aligned initiatives and road maps to understand the organization, communicate this understanding to stakeholders, and move the organization forward to its desired state", and Ross et al. [5] "Enterprise Architecture is the organising logic for business processes and information technology (IT) infrastructure reflecting the integration and standardization requirements of the company's model."

All these definitions introduce the main architectural components (processes, systems, technologies, components and their relationships) and covers methods to represent them, including both functional and non-functional requirements, by means of a set of views.

Enterprise Architecture provides various benefits, such as 1) Well-established solutions to architectural problems of organizations; 2) Help in documenting architectural design and implementation decisions; and 3) Facilitation of collaboration and communication between users.

A number of industry standard approaches have been proposed for defining enterprise architecture, such as the Zachman Framework for Enterprise Architecture [6] and The Open Group Architecture Framework (TOGAF) [7].

In this technological context, we are borrowing, adapting and refining the so popular and powerful patterns-oriented development to enterprise architectures. The following are some of enterprise architectures challenges that we are addressing specifically while adapting the pattern-oriented approach to TOGAF framework. Furthermore, for a novice designer or a software engineer who is not familiar with this mosaic of guidelines, it is hard to remember all design guidelines, let alone using them effectively.

In this paper, we introduced different categories of design patterns as a vehicle for capturing and reusing good analyses, designs and implementation applied to TOGAF framework.

5.2 BACKGROUND WORK

Introduced by the architect Christopher Alexander in 1977 [8], design pattern can viewed as a building block that we compose to create a design. A single pattern describes a problem, which appears constantly in our environment, and thus described the hart of the solution to this problem, in a way such as one can reuse this solution for different platform, without ever doing it twice in same manner [8]. For the cross-platform application development, patterns are interesting for three reasons; see also [9] for a more general discussion on patterns benefits:

- They come from experiments on good know-how and were not created artificially;
- They are a means of documenting architectures (out of building or software, enterprise in general);
- They make it possible in the case of a cross-platform development in team to have a common vision

Similar to the entire Enterprise Architecture community, the TOGAF community has been a forum for vigorous discussion on pattern languages for design, evaluation, and building a good architecture for the enterprises. The goals of the patterns is to share successful the design solutions among professionals and practitioners, and to provide a common ground for anyone involved in the design, development, enhanced usability testing, or the use of different systems. Several practitioners and designers have become interested in formulating various patterns of the same or different categories in the enterprise architecture destined to organizations

The idea of using patterns in TOGAF Framework is not new. Different pattern collections have been published including patterns for layout design [10-12], for navigation in a large information architecture as well as for visualizing and presenting information. In our work, we investigate categories of Patterns as a solution for cross-platform Enterprise Architecture and in particular, to solve the following design challenges

TOGAF [7] is an architecture framework that enables to design, evaluate, and build the right architecture for an organization. It is a mature Enterprise Architecture framework that is widely adopted by enterprises. TOGAF framework doesn't specify the architecture style—it is a generic framework TOGAF can be used in developing architecture. It consists of three main parts: The Enterprise Continuum, The TOGAF Resource Base and The TOGAF Architecture Development Method (ADM). ADM proposed a number of architectures shown and described below in Figure 1.

- Preliminary phase: This phase allows defining an Organization-Specific Architecture framework and the architecture principles. According the Dave Hamford [14], this phase is not a phase of architecture development;
- Phase A—Vision Architecture: This phase allows defining the scope of the foundation architecture effort, creating the vision architecture supporting requirements and constraints and obtaining approvals to proceed;
- Phase B—Business Architecture: This phase enables developing the detailed business architecture for analysing the gaps results;
- Phase C—Information System Architecture: This phase enables describing the Information Systems Architectures for an architecture project, including the development of Data and Application Architectures;
- Phase D—Technology Architecture: This phase enables developing a technology infrastructure that is used as a foundation for identifying all components that will support the development, implementation and deployment processes;
- Phase E—Opportunities and Solutions: This phase enables identifying opportunities and solutions and implementation constraints to deliver a more consistent architecture implementation;
- Phase F—Migration planning: This phase allows choosing and prioritizing all work packages, projects and to create, evolve and monitor the detailed implementation and migration plan providing necessary resources to enable the realization of the transition architectures;
- Phase G—Implementation Governance: This phase allows providing an architectural oversight of the implementation;
- Phase H—Architecture Change management: This phase allows establishing procedures for managing change to the new architecture;•
- Phase Requirement Management: This phase allows managing architecture requirements throughout the Architecture Development Method (ADM), i.e., defining a process whereby requirements for enterprise architecture are identified, stored, and fed into and out of the relevant ADM phases

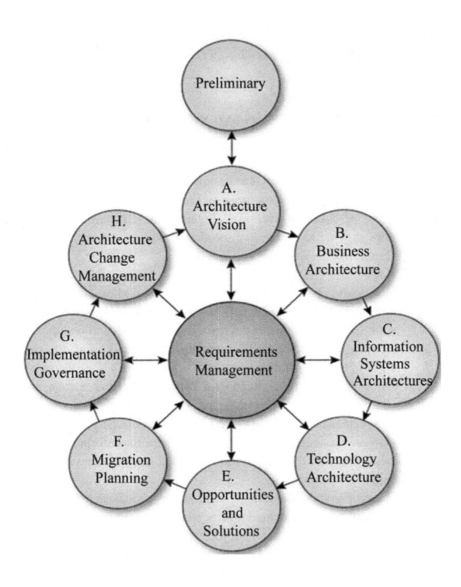

FIGURE 1: TOGAF framework [7]

By combining different categories of patterns, the professionals and experts can utilize pattern relationships and combine them in order to produce an effective and coherence design solution by using fully service-oriented approach that TOGAF has adopted. As a result, patterns become a more effective vehicle that supports design reuse and building organizational capabilities.

5.3 THE PROPOSED PATTERNS TAXONOMY FOR TOGAF FRAMEWORK

We propose at least ten categories of design patterns used to combine them to produce pattern-oriented enterprise architecture by applying the composition rules described in Section 4. Together, these patterns with their relationships provide an integrative solution to address the multifaces of TOGAF Framework (Figure 2):

1. Specification Patterns. This category of patterns allows understanding and clarifying the adopted strategy context, goals, and business architecture principles to the stakeholders in order to coordinate, and integrate the specifications of different activities at different levels of the organization.
2. Vision Patterns. This category of patterns describes a clear and stimulating vision of architecture to develop for addressing its requirements and constraints, and to meet the defined goals and objectives. These patterns communicate share the information with stakeholders on the signification of aimed goals by the vision and emphasize its importance.
3. Process Patterns. This category of patterns coordinates the actions and operations that related together, in serial or in parallel manner, in order to reach a common objective. The actions are the activities executed par human. The operations are the activities executed and controlled automatically by a software system. When a process is composed only with operations, then we called it an automated process.

4. Governance Patterns. This category of patterns describes the manner that all architectures of TOGAF framework are well-governed and managed successfully by taking into account and addressing both potential risks and potential value of the enterprise architecture. These patterns provide and inform the proper functioning of these various architectures, and specially their deployment and interaction. Theses architectures are linked by sequential interdependencies form. Indeed, they exchange together to produce the desired outcomes. Information must propagate between the involved architectures during the execution to harmonize their efforts to obtain better governance.

5. Migration Planning Patterns. This category of patterns describes and explains the important strategies of migration plan that were proven with execution. This effective plan consists of four key steps such as definition of needs, design, implementation, and tests. In addition, these patterns have to address the details of overall aspects through these strategies to ensure the optimal quality of the migrated functionalities of systems by including the best practices in order to develop the detail of the target organizational architecture.

6. Usability Patterns. This category of patterns focuses on dealing with the relationships between internal software attributes and externally visible usability factors and how these patterns can lead to a methodological framework for improving the "Opportunities and Solutions" architecture, and how these patterns can support the integration of usability in the software design process. In addition, these patterns expose knowledge that has been gained from different projects by many experts over many years.

7. Architecture Patterns. This category of patterns describes and gives information about the type of technological infrastructure to develop. Indeed, these patterns will support and enable the different business services implementation and deployment by using Service-Oriented Architecture (SOA) components of TOGAF framework.

8. Information Patterns. This category of patterns describes different conceptual models and architectures for organizing the underlying

content across multiple pages, servers and computers. Such patterns provide solutions to questions such as which information can be or should be presented on which device.

9. Business Patterns. This category of patterns describes a communication between the vision of organization with its business subjects with its objectives and its environment model such as actors, roles, and business service or functional or information or decomposition diagrams, business interaction, business footprint, product lifecycle diagram and all business processes involved.

10. Interoperability Patterns. This category of patterns is useful for decoupling the organization of these different categories of patterns as outlined in Figure 2, for the way information is presented to the user, and for the user who interacts with the information content. Patterns in this category generally describe the capability of different architectural programs to exchange data, via a common set of exchange formats considered as a service, to read and write under the same file formats, and to use the same protocols

Communication and interoperability patterns are useful for facilitating the mapping of a design between architectures of TOGAF framework

Gamma et al. [13] offer a large catalog of patterns for dealing with such problems. Examples of patterns applicable to interactive systems include: Adapter, Bridge, Builder, Decorator, Factory Method, Mediator, Memento, Prototype, Proxy, Singleton, State, Strategy, and Visitor.

5.4 PATTERN COMPOSITION RULES

A creation of an Enterprise Architecture pattern oriented design exploits several relationships between patterns. Based on previous work [15], we identify five types of relationships.

1. Similar is a relationship, which applies to the same category of patterns. Two patterns (X, Y) are similar, or equivalent, if, and only if, X and Y can be replaced by each other in a certain composition. This means that X and Y are patterns of the same category and

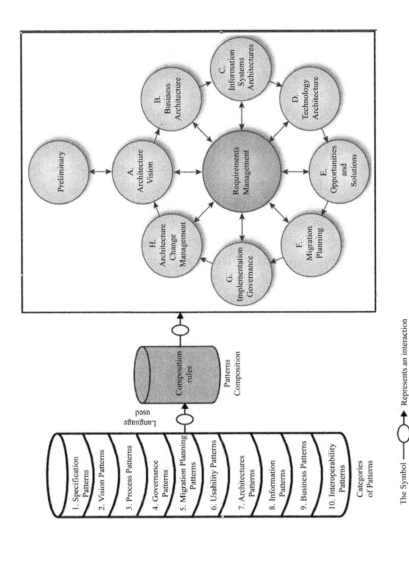

FIGURE 2. Pattern-oriented TOGAF framework

they provide different solutions to the same problem in the same context. For example, the Index Browsing and Menu Bar patterns are similar. They both provide navigational support in the context of a medium-sized.

2. Competitor is a relationship that applies to two patterns of the same patterns category. Two patterns (X, Y) are competitors if X and Y cannot be used at the same time for designing the same artifact relationship that applies to two patterns of the same pattern category. Two patterns are competitors if, and only if, they are similar and interchangeable. For example, the Web patterns Convenient Toolbar and Index Browsing are competitors. The Index Browsing pattern can be used as a shortcut toolbar that allows a user to directly access a set of common services from any interactive system. The Convenient Toolbar, which provides the same solution, is generally considered more appropriate.

3. Super-ordinate is the basic relationship to compose several patterns of different categories. A pattern X is a super-ordinate of pattern Y, which means that pattern Y is used as a building block to create pattern X. An example is the Home Page pattern, which is generally composed of several other patterns.

4. Subordinate. If pattern X is super-ordinate of Y and Z then Y and Z are sub-ordinate of X. This relationship is important in the mapping process of patternoriented design from an architecture to another one. For example, the Convenient Toolbar pattern is a sub-ordinate of the Home Page pattern for either a PDA or desktop platform. Implementations of this pattern are different for different devices.

5. Neighboring. Two patterns (X, Y) are neighboring if X and Y belong to the same pattern category. For example, the sequential and hierarchical patterns are neighboring because they belong to the same category of patterns, and neighboring patterns may include the set of patterns for designing a specific page such as a home page.

5.5 AN ILLUSTRATIVE EXAMPLE

This section describes the design patterns illustrating and clarifying the core ideas of the pattern-oriented approach and its practical relevance.

This case study illustrates how patterns are used to formalize and design the requirements of various architectures constituent TOGAF framework

In what follows, we have introduced some concrete examples of this mosaic of patterns that we have been using. These examples have shown also the need to combine several types of patterns to provide solutions to complex problems. The list of patterns is not exhaustive. There is no doubt that more patterns are still to be discovered, and that an endless number have yet to be invented

Interoperability patterns are fundamental patterns to facilitate the communication between requirements management phase and other architectures of TOGAF framework. Example of patterns that can be considered to ensure the interoperability of architectures include Adapter, Bridge, Builder, Decorator, Facade, Factory Method, Mediator, Memento, Prototype, Proxy, Singleton, State, Strategy, Visitor [13]

The Adapter pattern is very common, not only to remote client/server programming, but to any situation in which there is one class and it is desirable to reuse that class, but where the system interface does not match the class interface. Figure 3 illustrates how an adapter works. In this figure, the Client wants to invoke the method Request() in the Target interface. Since the Adaptee class has no Request() method, it is the job of the Adapter to convert the request to an available matching method. Here, the Adapter converts the method Request() call into the Adaptee method specificRequest() call. The Adapter performs this conversion for each method that needs adapting. This is also known as Wrappering.

5.6 DISCUSSION

The types of TOGAF architectures that are recommended for some the most popular patterns and which can be used to redesign of development systems for different architectures

In this paper, we have introduced a pattern-oriented design method that essentially exploits different categories of patterns. This approach is a significant improvement over non-structured migration methods currently in use, for the following reasons:

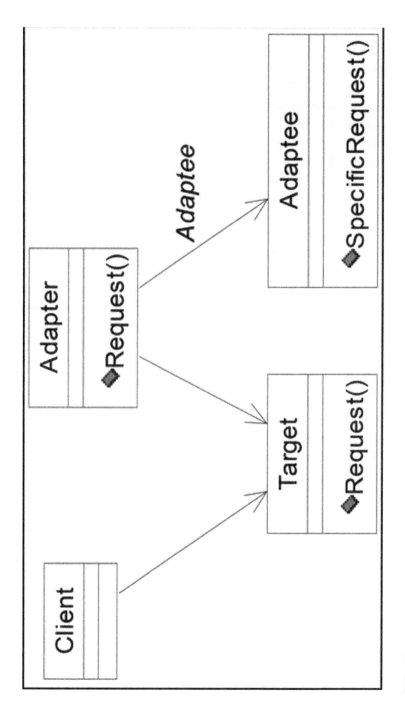

FIGURE 3: Adapter pattern.

- The method provides a standardized table of patterns, thereby reducing the redesign effort and ensuring consistency in redesign.
- The method helps designers in design choices associated with (1) the size of the source architecture and target architecture and (2) the amount of information to maintain in migrating from the source architecture to the target architecture
- The method is simple enough to be used easily by novice designers, as compared to reengineering which currently requires a considerable degree of expertise and abstract reasoning ability

Pattern-oriented approach offers the very useful ability of easily building multiple architecture-specific designs. However, the current state of the art in patterns and cross-architecture research is not yet mature enough to handle all the requirements of pattern-oriented design. More research must be addressed to define the multiple levels of abstraction of patterns and to create a clear, well-structured taxonomy of patterns. The simplified taxonomy presented in Section 3 is a starting point. Thus, within a pattern-oriented framework, the simplified "redesign and design" method proposed here is currently the most practical approach for migration of systems between architectures.

5.7 CONCLUSIONS

In this paper, we have identified and proposed ten categories of patterns, providing examples, for a patternoriented architecture for TOGAF framework to demonstrate when a pattern is applicable or required during the design process, how it can be reused and the underlying best practices to come up with reusable design solutions

Our experiences highlighted also that in order to render the patterns understandable by novice designers and engineers who are unfamiliar with enterprise architecture, patterns should be presented to developers using a flexible structure to represent patterns, to make it easy for both the pattern authors, reviewers and users

One of the major problems we can find is that mastering and applying several types of patterns require indepth knowledge of both the problems and forces at play and most importantly must ultimately put forth battletested solutions. As such, it is inconceivable that pattern hierarchies will evolve strictly from theoretical considerations. Practical research and industry feedback are crucial in determining how successful a pattern-oriented design framework is at solving real-world problems. It is therefore essential to build an "academia-industry bridge" by establishing formal communication channels between industrial specialists in patterns, enterprise architecture design patterns such as TOGAF framework as well as pattern researchers. Such collaboration will lead, at to a common terminology which essential making the large diversity of patterns accessible to common TOGAF framework designers

Future work will require the classification of each pattern and the illustration of each of them in UML class and sequence diagrams for each architecture of TOGAF framework. Next, some relationships will have to be defined between patterns so that they can be combined to create models based on the resulting patterns. Also, the design patterns need to be evaluated using different evaluation standards and methods and the formal descriptions of the proposed patterns using the formal language such as XML and its derivatives to increase the number of these formal descriptions which is also conducive to the future engineering application.

REFERENCES

1. Institute for Enterprise Architecture Developments, 2011. http://www.enterprise-architecture.info/
2. R. E. Giachetti, "Design of Enterprise Systems, Theory, Architecture, and Methods," CRC Press, Boca Raton, 2010.
3. P. Weill, "Innovating with Information Systems: What Do the Most Agile Firms in the World Do?" 6th e-Business Conference, Barcelona, 2007.
4. Enterprise Architecture Center of Excellence, 2011. http://eacoe.org/index.shtml
5. J. W. Ross, P. Weill and D. C. Robertson, "Enterprise Architecture as Strategy," Harvard Business Press, Boston, 2006.
6. J. A. Zachman, "A Framework for Information Systems Architecture," IBM Systems Journal, Vol. 26, No. 3, 1987, pp. 276-292. doi:10.1147/sj.263.0276
7. Open Group, 2008. http://pubs.opengroup.org/architecture/togaf9-doc/arch/index.html
8. C. Alexander, S. Ishikawa, M. Silverstein, M. Jacobson, I. Fiskdahl-King and S. Angel, "A Pattern Language," Oxford University Press, New York, 1977.
9. F. Buschmann, "What is a Pattern?" Object Expert, Vol. 1, No. 3, 1996, pp. 17-18.
10. J. Tidwell, "Common Ground: A Pattern Language for Human-Computer Interface Design," 1997. http://www.mit.edu/~jtidwell/common_ground.html

11. T. Coram and J. Lee, "Experiences—A Pattern Language for User Interface Design," 1998. http://www.maplefish.com/todd/papers/experiences
12. M. V. Welie, "The Amsterdam Collection of Patterns in User Interface Design," 1999. http://www.cs.vu.nl/~martijn/patterns/index.html
13. E. Gamma, R. Helm, R. Johnson and J. Vlissides, "Design Patterns: Elements of Reusable Object-Oriented Software," Addison-Wesley, Reading, 1995.
14. Dave Hornford, SOA/TOGAF Tutorial, 2010. https://www.opengroup.org/conference-live/uploads/40/22062/hornford.pdf
15. F. J. Budinsky, M. A. Finnie, J. M. Vlissides and P. S. Yu, "Automatic Code Generation from Design Patterns," IBM Systems Journal, Vol. 35, No. 2, 1996, pp. 151-171. doi:10.1147/sj.352.0151

CHAPTER 6

A NEW CONTENT FRAMEWORK AND METAMODEL FOR ENTERPRISE ARCHITECTURE AND IS STRATEGIC PLANNING

MOUHSINE LAKHDISSI and BOUCHAIB BOUNABAT

6.1 BACKGROUND

With the complexity of today's information systems and the necessity to make the existing IT assets more agile to provide for the constant business change, the task of governing and planning for IT assets become a key success factor for IT.

Information Systems Strategic Planning is the discipline that deals with this task. Unfortunately, it hasn't evolved with the same speed as other fields in the IT sphere. Most of the techniques, approaches and methods related to IT Strategic Planning date back to the '80s or '90s and are most often oriented business strategic planning rather than IT strategic planning [1]. As a matter of fact, they don't take into account the complexity of today's information system and their diversity. Furthermore, this field lacks from a formal, rigorous and agreed upon methodology and suffers from the absence of tools to support, structure and industrialize the discipline

This chapter was originally published under the Creative Commons License or equivalent. Lakhdiss M and Bounabat B. A New Content Framework and Metamodel for Enterprise Architecture and IS Strategic Planning. International Journal of Computer Science Issues *9,2 (2012), ISSN (On-line): 1694-0814.*

Enterprise Architecture is a really promising discipline aimed at capturing the as-is architecture of an enterprise, defining the target and the roadmap to get from existing to desired state. In that way, it is tightly related to ISSP and it can provide a framework to fill the gap and contribute in structuring and formalizing ISSP field. Enterprise Architecture benefits from a standardization effort as well as from tool support. Deliverables and artifacts are generally well defined and structured in the existing frameworks.

Existing Enterprise Architecture frameworks are of different types. While some frameworks like Zachman [2] define a taxonomy for architecture artifacts, others like TOGAF [3], tend to describe a process to produce architecture deliverables [4]. The main concept underlying both the process and the taxonomy is the metamodel to describe architecture elements and to produce architecture deliverables.

This metamodel is often either very poor to describe fully the architecture or not well structured to define the dependencies and the relationships between elements. We think that in order to define a more rigorous and structured methodology for ISSP, it is necessary to define a rich and structured metamodel covering both architecture elements (processes, applications, data..) and transformation elements (programs, projects, budgets). This metamodel is the main focus of our work aimed at defining a new methodology for ISSP.

Our aim in this paper is to demonstrate the insufficiencies, deficiencies and inconsistencies in existing ISSP methods and show how a new methodology based in part on the Entreprise Architecture practice could be proposed to address these problems. The metamodel we project to define could be used as a platform for describing the architecture, evaluating it and defining the needed transformations and planning them in term of programs/projects.

The second section presents ISSP and EA, compares the two disciplines and tries to bridge the gap between them. The content framework and the underlying metamodel is introduced in the the third section as a way to combine ISSP and EA. The fourth section describes the suggested metamodel and content framework. A comparison is made in the fifth section with existing metamodels before presenting future work and directions.

6.2 IT STRATEGIC PLANNING AND ENTERPRISE ARCHITECTURE

6.2.1 IT STRATEGIC PLANNING

Strategy is defined by Chandler as "The determination of the basic long-term goals and objectives of an enterprise and the adoption of courses of action and the allocation of resources necessary for carrying out these goals" [6] and by Porter as "The art to build durable and defendable competitive advantage" [7].

One of the most complete definitions was given by [8], "A fundamental framework for an organization to assert its vital continuity, while, at the same time, forcefully facilitating its adaptation to a changing environment."

For most of the definition, strategic planning is focused on three main questions:

- Where we are?
- Where we want to go?
- How to get there?

IS Strategic planning has been defined by [9] as the process of identifying a portfolio of applications/projects that can help an organization achieve its business strategy. Its focus is on defining the IT roadmap in term of key initiatives, projects and transformations to be made on the existing information system with two main intentions:

- How to align information systems with business needs and overall strategy?
- How to use information technology to change and impact the business?

Due to the complexity of today's information systems and the diversity of enterprise's technology approaches, many methods have been defined to structure the ISSP process and techniques have been defined to address some aspects of the discipline. [10] classifies ISSP methods into two categories:

- Impact methods: trying to make It help create a positive impact and drive the change of the business

- Alignment methods: where the main focus is on aligning IT to respond to business needs and to help achieve strategic goals

Among the methods used in IT Strategic Planning we can cite Critical Success Factors (CSF) [1] which could be considered as an impact and alignment method, Business Systems Planning (BSP) [1], Porter's Value Chain [7], and Scenarios [1]. Methods can be grouped together to constitute a methodology. Methodologies used for ISSP include those of the CCTA (12) and Boar (13).

Many IT vendors and consultancy organizations use proprietary methods and/or methodologies, some of which are adaptations of open source approaches. Examples are Arthur Andersen's Method/1 and Coopers and Lybrand's Summit [9]. It is also well known that organizations often develop their own in-house methodologies, often based on open or proprietary methods or approaches [9]

6.2.2 ENTERPRISE ARCHITECTURE

ISO/IEC 42010: 2007 defines "architecture" as: "the fundamental organization of a system, embodied in its components, their relationships to each other and the environment, and the principles governing its design and evolution." The Open Group defines it as [3]: "A formal description of a system, or a detailed plan of the system at component level to guide its implementation. The structure of components, their inter-relationships, and the principles and guidelines governing their design and evolution over time"

An architecture is typically made up of:

- a picture of the current state
- a blueprint, vision or detailed description for the future
- a road-map on how to get there

Enterprise Architecture appeared in the eighties thanks to John Zachman who introduced the framework that bears his name. This framework consists of taxonomy for producing architecture artifacts from different viewpoints and perspectives. As a matter of fact, Enterprise Architecture has been defined by Zachman [2] as a "set of descriptive representations (i.e. 'models') that are relevant for describing an Enterprise such that it can be produced to management's requirements (quality) and maintained over the period of its useful life".

Several other frameworks appeared subsequently, most of them initiated by government bodies like TAFIM (Technical Architecture Framework for Information Management), DODAF, MODAF or FEAF especially due to the requirement of the Clinger-Cohen Act.

IT consulting firms created their own EA frameworks, based on the feedback from projects they undertook. Gartner as well as Cap Gemini or Accenture have their own EA frameworks which could be more accurately considered as EA practices as stated by [4].

The Open Group Architecture Framework (TOGAF) started with TA-FIM and reproduced practices and techniques used in other framework to constitute an EA framework of reference in the IT industry. TOGAF is with Zachman the two most used EA frameworks according to [5]. TOGAF consist of:

- A architecture development methodology describing the process
- A set of guidelines and techniques supporting the methodology
- A content framework with a metamodel describing the products (deliverables)
- Reference models that provide best practices to compare with
- A structure and description of the architecture repository (enterprise continuum)
- A capability framework for architecture governance and implementation

TOGAF could be used in combination with Zachman where TOGAF defines the process and Zachman the deliverables. Archimate [11] defines a notation for architecture elements but also defines its own metamodel for architecture description.

Enterprise Architecture could be used for different needs and in various contexts. It can operate as:

- A method to describe the enterprise as a whole with different levels and views of enterprise elements and their relationships. In this way it relates to Enterprise Modeling as was stated by Lillehagen et al.[17]
- A way to align the IS environment with the business reality and the strategic goals or to assess this alignment as described by Bounabat[18] and Elhari[19]

- A modelling structure to define the vision for IS evolution or to describe in detail the IS to-be state
- A process to plan the migration between the as-is situation and the to-be state.

All these EA use cases could be related to a step or a phase in term of process or deliverables of ISSP.

6.2.3 COMPARISON AND CORRELATION

A theoretical comparison of IT Strategic Planning and Enterprise Architecture was conducted by Wilton [20] and Beveridge and Perks [21]. These comparisons concluded that both ISSP and EA share the same intent and scope. Wilton [22] gave a more empirical comparison based on a survey which led to establishing a significant correlation between the two activities in term of topics they cover.

The main difference that was highlighted by Wilton [22] is that ISSP tends to be process-oriented with little specification of deliverables and content while EA is product-oriented in that it defines the way the as-is and to-be state are described and modelled.

We think that this difference tends to disappear due to the progress made in the field of Enterprise Architecture. As a matter of fact, with frameworks like TOGAF the gap is being bridged with a detailed process to produce architecture deliverables.

Furthermore, we think that other differences are to be considered. One of the main differences that still exists and that is related to the Enterprise Architecture practice is the fact that there is no concrete link between the architecture description and the programs/projects defined in the roadmap. This lack of correlation makes it difficult to address the strategic planning main objective which is planning for IT transformations with existing Enterprise Architecture frameworks.

6.2.4 BRIDGING CONCEPT: TRANSFORMATION

A project is defined by (PMBoK) as "A project is a temporary endeavor undertaken to create a unique product, service or result."

This definition underlines the fact that a project is intended to create a product/service or result. It doesn't mention the elements that the project will impact whether they are new elements created or existing element transformed.

A project—in the context of ISSP and EA—could be defined as a set of transformations (including creations) applied on architecture elements. These elements could be business elements, application elements, data elements or technical elements or a combination of them.

Elements are the basic constituents of architecture like applications, processes, servers, databases...etc. These elements are combined to create architecture models and diagrams. The transformations of these elements are combined as well to create ISSP's projects and programs.

6.3 IMPORTANCE OF A FRAMEWORK AND METAMODEL

The ISO/IEC 42010: 2007 definition of architecture as "the fundamental organization of a system, embodied in its components, their relationships to each other and the environment, and the principles governing its design and evolution." highlight unequivocally the importance of the organization of elements and their relationships. This structure is defined through a metamodel of architecture elements.

Enterprise Architecture is supposed to produce architecture artifacts. These artifacts are based on an architecture content framework as defined by TOGAF or an architecture map.

6.3.1 CONTENT FRAMEWORK

The content framework defines the layers, views, questions and aspects that architecture description deals with. The importance of this framework is that it organizes, classifies and links architecture elements and artifacts. It is also interesting because it ensure the coherence and exhaustively of the metamodel.

The content framework is classically defined as a bidimensional grid with lines representing layers or views and columns representing concerns and classifications. The content framework defines elements of the metamodel in a high level way emphasizing the global structure rather than the detailed model.

6.3.2 METAMODEL

The metamodel is the backbone of architecture description and methodology. The metamodel guarantees the exhaustiveness of overall architecture work and the coherence and alignment of architecture layers.

It is similar in form to a Conceptual Data Model or a Class Diagram in UML. It is important in term of objects definition, attributes definition and relationships.

- Objects definition ensures the exhaustiveness and coverage of aspects as standardization and integration.
- Attributes provide the way to perform diagnosis and analysis on existing and future assets. Attribute can also cover aspects like security and performance necessary to the evaluation process.
- Relationships are very important to perform Gap Analysis inside the same layer and for alignment needs between layers.

6.4 EXISTING CONTENT FRAMEWORKS AND METAMODELS

Many metamodels have been defined explicitly or implicitly by EA frameworks. They are of different natures and focus depending on their intent. Some of them are poor in term of business or IS content. Others don't take into account some aspect tightly related to EA and ISSP like:

- Requirements
- Strategy
- Standards
- Program and projects

We described in our paper [23] each metamodel (TOGAF, Zachman, Archimate, EA Tools) with a critical view of each one. The summary of this analysis is presented in table 2.

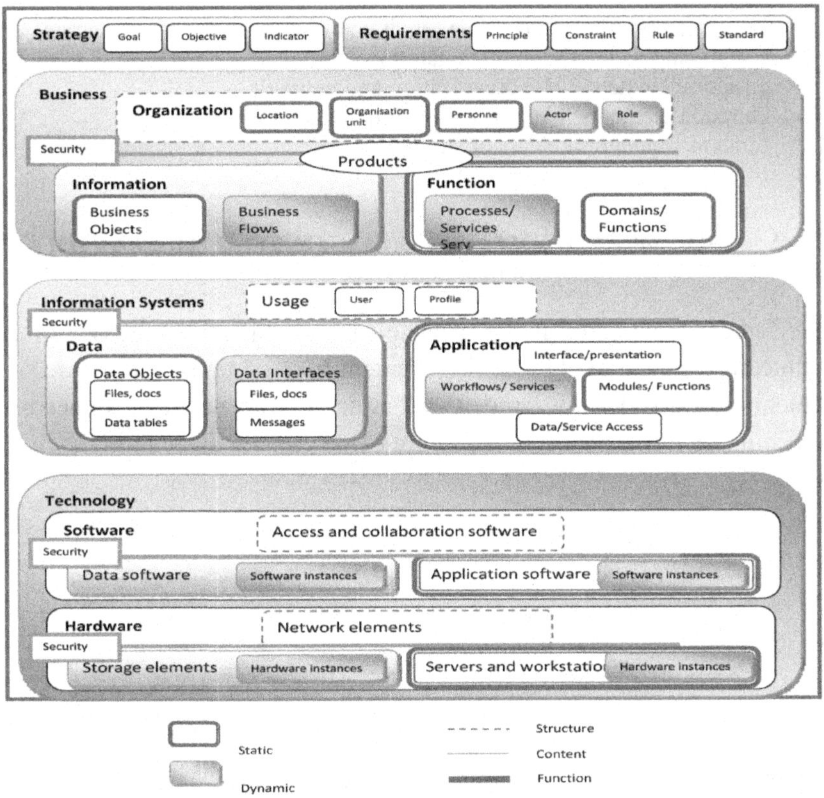

FIGURE 1: New content framework: Neoxia Architecture Map (NAM)

6.5 PRESENTATION OF A NEW CONTENT FRAMEWORK AND METAMODEL

6.5.1 NEW CONTENT FRAMEWORK : NEOXIA ARCHITECTURE MAP

We introduce here a new content framework that is based on feedback from consulting projects on EA and ISSP. We call it NEOXIA Architecture Map. In this content framework we differentiate between

- Static element: tending to describe an element in a static most often hierarchical way
- Dynamic element: focusing on the dynamic view of the same element We also distinguish between three natures of element:
- Structure elements: like organization and network
- Function elements: like services or functions
- Content element: like data or storage

6.5.2 NEW METAMODEL: NEOXIA CONTENT METAMODEL

The content metamodel is the mechanism by which we suggest to mix architecture and strategic planning element base on transformations. The content metamodel follows the overall structure of the content framework and could be illustrated as in the figure below.

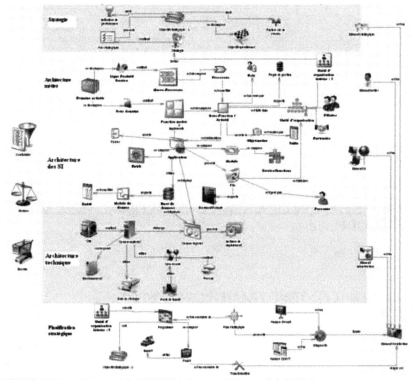

FIGURE 2: New Metamodel : NEOXIA Content Metamodel (NCM)

The suggested metamodel is composed of five layers:

- Strategy
- Business
- Information Systems
- Technology
- Strategic planning

All layers are interrelated with static and dynamic element of the three natures: function, structure and content. Every layer is connected with the layer below with a realization link. A process is automated in an application which uses a database and are both deployed in a server. This dependence is fundamental to align the IS with the Business Architecture and the Technology with the IS Architecture. This link allows us also to analyze the gap between layers in term of coverage to make it possible to fill this gap in the strategic plan.

The metamodel could be also represented as package and class diagrams. We focus in Figure 3 on the strategic planning layer.

The central concept is "Transformation" which is a migration from an as-is state to a to-be state of an architecture element. An architecture element could be any architecture object of the metamodel (ex: process, application, Hardware server…etc). A transformation is operated either:

- A realization of a strategic objective: this allows us to align the to-be IS situation with the strategy and to justify the strategic plan investments
- A consequence of an IT or business requirement (principle, standard, rule, constraint) defined by the organization
- A result of a gap analysis: in that case the gap is observed on one or more architecture elements and the transformation is a way to fill the gap.

We have as a result three types of transformations :

- Alignment transformation
- Requirement transformation
- Gap transformation

Many transformations are grouped into projects which are managed through programs.

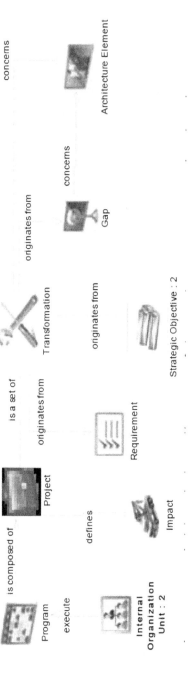

FIGURE 3: Highlevel class diagram of the package "IT Strategic Planning"

6.6 COMPARISON WITH EXISTING CONTENT FRAMEWORK AND META MODELS

The suggested metamodel defines the detailed structure and relations of architecture elements starting from strategy and requirement and going through the different levels (Business, Information Systems and Infrastructure) with the necessary link with Strategic Planning element like gap analysis, program and project.

In addition of giving a more detailed structure for IS and infrastructure levels which are often poorly defined, the main contribution of this metamodel is the link it establishes between architecture elements and strategic planning elements.

We summarize a theoretical comparison of our metamodel with existing metamodels presented in section 4.

TABLE 2: Comparison of the new metamodel with existing ones

	New metamodel NCM	TOGAF	Archimate	Zachman	EA tools metamodels
Requirement	Yes	No	No.	Undergoing	Partially Yes
Strategy	Yes	Yes	No	Yes	No
IT Planning	Yes	No	No	No	Yes to some extent
Link between projects and architecture	Yes	No	No	No	No
Standards	Yes	No	No	No	Yes. (Not native)
Strategy definition	Yes	Yes	No	Yes	Yes
Business description	Detailed	Detailed	Detailed	Average	Poor
IS description	Detailed	Poor	Average	Detailed	Detailed (Depending on tool)
Infrastructure description	Detailed	Poor	Average	Detailed	Detailed (Depending on tool)
Tool support	No	Partial	Yes	Partial	Yes
Methodology support	No, Undergoing	Yes	Yes	No	No
Independence	Yes	Yes	Yes	Yes	No

6.7 CASE STUDY AND IMPLEMENTATION

The framework and metamodel described above were used in a IT Strategic Plan in the financial market regulation agency to describe as-is and to-be IT state and to define the migration plan in term of projects and transformations. The IS Strategic Plan was part of an e-government strategy to make all interaction between stakeholders in the market based on internet and Electronic Data Interchange which make the underlying architecture complex enough to provide for a good testing environment.

The whole existing and future information systems component and architecture were described and modeled based on the content framework described above and using the proposed metamodel. This description leaded to a thorough visibility on existing and future state for all stakeholders of the project which was necessary to take the right decisions concerning the evolution scenarios. Moreover, a program of project was defined with for each project a set of transformations of architecture element from an existing to a future state. This was very beneficial for impact analysis, projects dependencies and load estimation of the projects.

6.7.1 IMPLEMENTATION

The implementation in our context has a double goal:

- To make sure the metamodel is realist and feasible
- To constitute a platform for a future ISSP tool (which is a much needed tool in IT Governance) The idea is to build a tool that makes it possible to :
- Describe graphically and in term of properties all architecture elements of our metamodel
- Define the dependencies and links between these element based on the metamodel
- Store all elements and their dependencies in a repository
- Generate inventories, matrices and reports from the repository

Two scenarios of implementation were possible:

- Customize the metamodel and content framerwork of an existing EA tool
- Implement a new tool probably based on an open source existing one

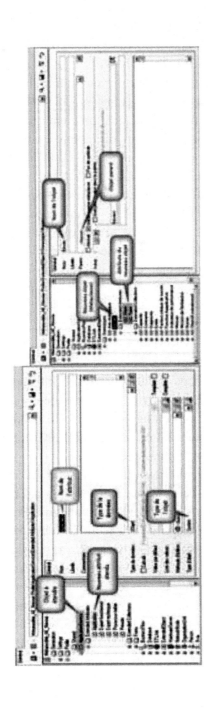

FIGURE 4: New Metamodel : NEOXIA Content Metamodel (NCM)

We explored both options with the e-Government Regulation Agen-cy case study which allowed us to measure the degree relevance of the metamodel.

The first option was carried out through a customization of Sybase Power AMC aimed at adapting its metamodel to NCM. Even if the tool is quite flexible, we have encountered a problem concerning the predefined objects of Power AMC which are difficult to customize as well as some dependencies which are hard to implement. Figure 4 shows some screen-shots of the customization screens.

We then explored building our specific tool based on the Eclipse Plat-form (Eclipse RCP, EMF and GEF). The tool is based on an XML storage of the model (values of attributes of objects). In Figure 5, are displayed some screenshots describing how we can create an architecture element with its graphical representation and its properties.

6.8 CONCLUSION AND FUTURE WORK

We think that based on this proposed metamodel, a new methodology could be defined to cope with the needs of ISSP and to complement and enrich existing EA metamodels. The metamodel described was already used successfully in consulting projects in the public and private sector and was able to capture more meticulously architecture element and to support the process of IS Strategic Planning.

The metamodel could be enriched to highlight crossover architecture aspects like security, performance and integration. These aspects are very important in evaluating existing IT assets and in defining their target state.

A planned continuation of this work is to continue on developing a basic modeling tool (or adapt an existing one) based on this metamodel and content framework with support to Enterprise Architecture as well as IS Strategic Planning techniques and activities. The tool will allow to put into practice the Metamodel and to demonstrate the added-value of the methodology.

Another extension is to formalize diagnosis and evaluation techniques into the meta model to make sure the whole IS Strategic Planning process is automated.

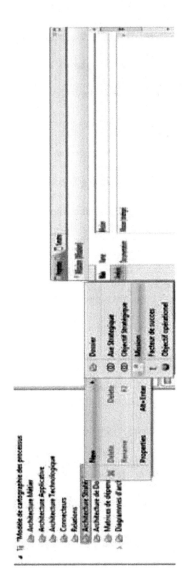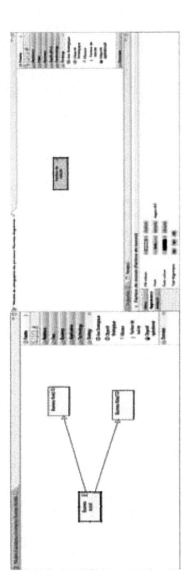

FIGURE 5: New Metamodel : NEOXIA Content Metamodel (NCM)

REFERENCES

1. Pant S., Hsu C.: Strategic Information Systems Planning: A Review, 1995 Information Resources Management Association International Conference.
2. Zachman J. : A framework for Information Architecture, IBM Systems Journal, 38, 2&3, 1987.
3. The Open Group .: "TOGAF as an Enterprise Architecture Framework", http://www.opengroup.org/architecture/togaf8-doc/arch/
4. Sessions R. : "A Comparison of the Top Four Enterprise-Architecture Methodologies", http://msdn.microsoft.com/enus/library/bb466232.aspx (2007)
5. Schekkerman J. : Trends in Enterprise Architecture 2005: How are Organizations Progressing?
6. Chandler, Alfred A., Jr. (1962), Strategy and Structure: Chapters in the History ofAmerican Industrial Enterprise, The MIT Press, Cambridge, MA
7. Porter, M. E. (1985). Competitive Advantage: Creating and Sustaining Superior Performance, Collier Macmillan, New York, N.Y.
8. Arnoldo C. Hax (1996), "Strategy Concept and Process: A Pragmatic Approach", Barnes and Noble, bn.com
9. Lederer, A. L. and Sethi, V. (1988). The Implementation of Information Systems Planning Methodologies, MIS Quarterly, September 1988, 445- 461.
10. Vitale, M., Ives, B. and Beath, C. (1986), "Identifying Strategic Information Systems," Proc. 7th Int'l Conf. Inf. Sys., San Diego, December 1986, pp. 265-276.
11. Archimate (2009), "Archimate 1.0 Specifications," http://www.opengroup.org/architecture/togaf8-doc/arch/
12. CCTA (1999). IS Strategy: process and products, Format Publishing Limited, Norwich.
13. Boar, B. (2001). The Art of Strategic Planning for Information Technology, John Wiley and Sons, New York, NY.
14. DODAF, U.S. DoD (2003). DoD Architectural Framework Version 1.0,
15. http://www.teao.saic.com/jfcom/ier/documents/DOD_architecture_framework_volume1.doc,
16. Lillehagen F., Karlsen D., Enterprise Architectures – Survey of Practices and Initiatives, Rapport Computas 2005.
17. Bounabat, B. (2006), "Enterprise Architecture Based Metrics for Assessing IT Strategic Alignment", European Conference On Information Technology Evaluation (ECITE), 2006.
18. Elhari K., Bounabat. B (2010), "Strategic Alignement Assessement Based on Enterprise Architecture", ICIME 2010
19. FEAF, 1999, "Federal Enterprise Architecture Framework Specification", http://www.whitehouse.gov/omb/e-gov/fea/
20. Wilton, D. (2001). The Relationship Between IT Strategic Planning and Enterprise Architectural Practice, Journal of Battlefield Technology, 1, 18-22.
21. Perks, C. and Beveridge, T. (2003). Guide to Enterprise IT Architecture, Springer-Verlag, New York.

22. Wilton, D. (2007). The Relationship between IS Strategic Planning and Enterprise Architectural Practice: a Study in NZ Enterprises, Information Resources Management Association (IRMA), Vancouver.
23. Lakhdissi M, Bounabat B (2011). "Toward a novel methodology for IT Strategic Planning". Proceeding of ICIME 2011, p 277-287

There is one table that does not appear in this version of the article. To view it, please visit the original version of the article as cited on the front page of this chapter.

CHAPTER 7

AGENT-ORIENTED ENTERPRISE ARCHITECTURE: NEW APPROACH FOR ENTERPRISE ARCHITECTURE

BABAK DARVISH ROUHANI and FATEMEH NIKPAY

7.1 INTRODUCTION

Enterprise architecture is a new approach to aligning business and IT within an organization for competitiveness. Enterprise Architecture is a comprehensive system that encompasses all activities aspects of an organization [12]. The exact and specific relationship among the components of the organization's architecture is an Enterprise Architecture's advantage, which is implemented by the Enterprise Architecture's program. In other words, the duty of enterprise architecture is to implement the enterprise architecture's structure in an organization. Comprehensive coverage of an organization's activities, causes enterprise architecture structure seems complex and ambiguous, so to avoid problems and to identify a suitable model, existence of a framework in enterprise architecture is vital. Utilizing a suitable framework facilitates the analysis of organization structure in order to determine the current status, optimal conditions and also defining the transfer functions [2][13].

This chapter was originally published under the Creative Commons License or equivalent. Rouhani BD and Nikpay F. Liver and Muscle in Morbid Obesity: Agent-Oriented Enterprise Architecture: New Approach for Enterprise Architecture. International Journal of Computer Science Issues, *9,6 (2012),* ISSN (Online): 1694-0814.

Different attitudes in enterprise architecture are established since EA is introduced by John Zackman. After a while, the framework provided by Zackman was separated into more specialized framework such as federal, financial and military. Each of these frameworks has the capability to cover the organization's activities in their own professional field. Combination of Service Oriented Architecture, Enterprise Architecture and Agile Architecture creates Enterprise Architecture based on services and Agile Enterprise Architecture. All these efforts were made to augment EA and increase success rate of EA's programs in an organization.

Since software architecture has a very constructive role in the successful implementation of enterprise architecture, so it's vital to utilize the new software architecture for the enterprise architecture which has already been tried by service-oriented architecture and Agile Architecture. Nowadays Agent-oriented architecture which has obtained its own special place in all software's aspects including analysis, design and implement, can obtain great success in projects with huge and complex structure [2][5][6].

7.2 AGENT-ORIENTED ARCHITECTURE

Agent-oriented architecture is formed based on the fact 'agent', which has the capability of autonomy in decision making, team work, work passively and being goal oriented.

These characteristics form the software operate dynamically and make appropriate decisions based on common interaction with each other in case of each event and then take appropriate reaction. Some agent-based system's applications are as follows:

- Solving problems that are inherently large and complex and requires a distributed mechanism for resolving them.
- Reduce processing costs (utilizing a large numbers of inexpensive processors is better than an expensive and powerful ones).
- To provide interactive between Legacies Legacy systems.
- Applications where their focus is on scalability.
- Providing solutions for problems that are inherently distributed. For example: Workflow Management, air Traffic Control

As it is obvious in agent definition an agent has the ability to perform an activity encapsulated in a flexible and independent environment in order to fulfill design goals. The environment is a place where the agents interact with each other and resolve operational and information requirements of each other's [7][8].

7.3 EXISTING ENTERPRISE ARCHITECTURE PROBLEMS

Most organizations during implementation enterprise architecture are faced with below problems [2]:

- Deflected and scattered focuses
- Project teams, are not familiar with existing enterprise architecture.
- Project teams; do not follow the enterprise architecture.
- Project teams; do not cooperate with the enterprise architects.
- The architectures are obsolete
- Less attention to architectural models
- Non-routine programs
- The tendency to do all this extra work just because it is good for the organization.

There are various theories for software development and each of them has been tested. The important point to mention is that agent-oriented architecture has the capability to solve all these problems. Agent-oriented architecture can deal with obstacles and solved problems from the beginning to the end of a project by defining agents as independent elements with ability to perform proper interaction with each other. The agent-oriented architecture has capabilities to focus on the business architecture.

7.4 PROPOSED APPROACH

By considering these points:

- The agents are the existence of solutions that have well-defined intervals.
- Agents operate in special environments and sense inputs of the environment's state through their sensors and operate on that environment by their effectors.

- Agents are designed for specific roles.
- Agents are independent and have control over the interior situation and their own behavior.
- Agents are able to provide flexible solution. It is necessary to be reactive to what happens in the environment surrounding, which helps them achieve their goals. It is necessary to have very active agents in order to get initial values and data to fulfill new goals.

Agent-oriented architecture has capabilities to solve current problems of enterprise architecture by its characteristics [7][8].

7.4.1 WHY AGENT-ORIENTED?

As EA makes a revolution in business and information of an enterprise, so it's vital to utilize an appropriate method in order to reduce complexity and manage this complex organization better. In this part the reasons are stated that why agent oriented architecture is a solution for EA's problems.

- Agent-oriented analysis is an appropriate method to divide the problem in complex systems.
- The main agent oriented abstracts is a useful tool in modeling complex systems.
- Agent oriented viewpoint is appropriate for detection and communication management of the organization. It is suitable for dependency management and the interaction exists in a complex system.

For example, if financial sub-system faces changes due to tax and legal laws, in a centralized and complex system all other sub-systems are affected and in many cases cause improper behavior in system. If the system is formed based on agent oriented capabilities, these kinds of problems rarely occurs, and active agents reconstruct overall system structure by means of defined structure automatically or in team activities. The system can overcome the problems of non-pre-defined changes, by its dynamic structure [4].

7.4.2 AGENT-ORIENTED ENTERPRISE ARCHITECTURE

In the complex system, relations between systems components are changed continuously, thus they require a set of components to acts as a conceptual unit when they are observed at different levels of abstraction. This view is quite compatible with experience of agent oriented viewpoints, by this definition it's obvious that facilities are provided for clear organizational communication.

There are various types of organizational communication in complex systems which are important for two reasons: First, a number of separate components can form a group together, secondly, they make it possible to fully explain the existence of high levels connections and defined their details (Figure 1) [9]. The effect of organizational communication and structures on systems' behavior, illustrate the importance of a flexible management .these communication are continuously changing and ability to adapt dynamically to new circumstances is often necessary. As it was mentioned before the first entity are the existence in agent oriented systems. So clear structures and flexible mechanisms of are the centrality of agent oriented patterns.

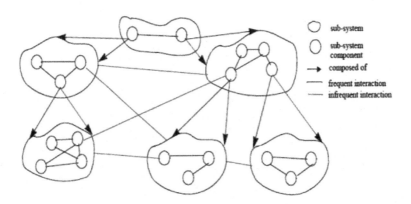

FIGURE 1: sample of complexity structure in complexity systems

If this ability is combined with computing mechanism, it would enable agent oriented systems to change according to their requirements and also, make many middle forms which are necessary to produce complex systems more rapidly. This means that agents or organizational groups can be produced relatively independent of each other, and this would ensure appropriate growth and proper coordination.

When the project teams works with assumption that they can do anything they want, and utilize any technology to make change in results, the result is that the works and data are duplicated and reusing them rarely happens. Systems won't work properly and they are incompatible with each other and cause each other to fail and increase costs dramatically.

A bad reality that exists is that almost a few number of software systems operate in a close environment while they should communicate with several and sometimes hundred other systems. In an organization, application must be effectively cooperate with other systems, as a result the application must be developed a little in order to prevent harmful effects on other systems .also they should be produced ideally to take advantage of systems and increase shared infrastructure. Each system must be constructed so it can fit in the existing environment of an organization, and it's better to represents future viewpoint in it. Such information should be registered in the enterprise architecture, current and future models in respect. The purpose of agent oriented EA is to ensure proper implementation of activities and providing dynamic structure in implemented system .to sum up agent oriented EA is result oriented and make the activities successful by utilizing clear definition of an agent's goals and providing an appropriate environment.

Agent oriented enterprise architecture is a method to define all aspects and different viewpoints of a busy and complex organization where unpredictable change in mission and technology are thoroughly impressive. These changes are unpredictable so the organization has not the ability to create and develop a specific plan to deal with them. Thus, agent-oriented enterprise architecture is capable to define organization's current situation and desire state by means of agent-oriented models, techniques and methods. Formulation and implementation of transfer pattern is one of enterprise architecture principles and is planned well in agent-oriented enterprise architecture. According to Agents abilities such as autonomy and interaction they can capable to cover all aspect of Enterprise Architecture. For reaching to this purpose define some items as well are very important (figure 2).

Those items are [9]:

- Defining accurate information/activity environment
- Determining accurate relation between enterprise systems (inter systems)
- Determining accurate performance and interaction between agents
- Determining accurate agent activities domain

7.5 CONCLUSIONS

Utilizing agent-oriented enterprise architecture is appropriate for organizations with complex missions and where missions are highly dependent on each other. Generally, by means of agent-oriented architecture's patterns, in Enterprise Architecture, an organization will be able to perform the following:

- Increasing the likelihood of successful implementation of Enterprise Architecture
- Changes in the organization's plans
- Increase competitiveness against other organizations.

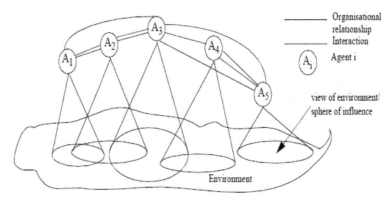

FIGURE 2: Focal vision of multi agent system

- Development of the lateral structures through proper implementation of Enterprise Architecture
- Development of long term and medium term organizational goals
- The development of enterprise data structures
- Stability in the development of enterprise plans

REFERENCES

1. Babak Darvish-Rouhani, Sadegh Kharazmi, " Presenting New Solution Base On Business Architecture For Enterprise Architecture", IJCSI International Journal of Computer Science Issues, Vol. 9, Issue 3, No 1, May 2012, ISSN (Online): 1694-0814

2. Babak Darvish Rouhani, Hossain Shirazi, Ali Farahmand Nejad, Sadegh Kharazmi. "Presenting a Framework for Agile Enterprise Architecture", 1st International Conference of IEEE Information Technology 2008 (IT2008), Gdańsk, Polnad.

3. Yan Zhao, Ph.D., Director, Enterprise Architecture, CGI Federal ; "Enterprise Service Oriented Architecture (ESOA) Adoption Reference"IEEE International Conference on Services Computing (SCC'06)-2006

4. N. R. Jennings, K. Sycara and M. Wooldridge (1998) "A Roadmap of Agent Research and Development" Int Journal of Autonomous Agents and Multi-Agent Systems 1 (1) 7-38.

5. Namkyu Lim, Tae-gong Lee, Sang-gun Park; "A Comparative Analysis of Enterprise Architecture Frameworks based on EA Quality Attributes"; 10th ACIS International Conference on Software Engineering, Artificial Intelligences, Networking and Parallel/Distributed Computing; 2009

6. Gonc¸alo Antunes, Jos´e Barateiro, Christoph Becker, Jos´e Borbinha, Ricardo Vieira;" Modeling Contextual Concerns in Enterprise Architecture"; 15th IEEE International Enterprise Distributed Object Computing Conference Workshops, 2011

7. Jaelson Castro, Manuel Kolp, John Mylopoulos; "DEVELOPING AGENT-ORIENTED INFORMATION SYSTEMS FOR THE ENTERPRISE"; Proceedings Second International Conference On Enterprise Information Systems, Stafford, UK, July 4-7, 2000

8. LIU Xiang; "A multi-agent-based architecture for enterprise customer and supplier cooperation context-aware information systems"; Third International Conference on Autonomic and Autonomous Systems (ICAS'07), 2007

9. M. Wooldridge (1997) "Agent-based software engineering" IEE Proc. on Software Engineering, 144 (1) 26-37.

10. A. Wegmann, On the Systemic Enterprise Architecture Methodology (SEAM), in: International Conference on Enterprise Information Systems (ICEIS), 2003.

11. Institute of Electrical and Electronics Engineers, IEEE STD 1471, Recommended Practice for Architectural Description of Software-Intensive Systems, The Institute of Electrical and Electronics Engineers, Inc., New York, New York, 2000.
12. J. Schekkerman, "How to Survive in the Jungle of Enterprise Architecture Frameworks," Trafford, 2004.
13. J.Zachman, A framework for information systems architecture, IBM Systems Journal 26 (3) (1987).

CHAPTER 8

ADaPPT: ENTERPRISE ARCHITECTURE THINKING FOR INFORMATION SYSTEMS DEVELOPMENT

HANIFA SHAH and PAUL GOLDER

8.1 INTRODUCTION

Much progress has been made in recent years in developing structures to describe the enterprise and to facilitate the development of information systems that appropriately complement the strategy of the enterprise. Despite the success of the enterprise architecture approach there are still major problems in achieving organisational change and in driving the re-alignment of IT systems. The complexity of modern organisations in terms of the business, legal and technological environment demands an architectural approach. Businesses are faced with ongoing and continual change to which they must respond in order to ensure success and even survival. The increasingly competitive environment demands a customer-focused approach. All of these factors contribute to the complexity and uncertainty faced by organisations resulting in an inability to be appropriately responsive to both internal and external events. Underlying this complexity and uncertainty is the gap between an organisation's business objectives and its underlying IT infrastructure. There is a need for information that is

This chapter was originally published under the Creative Commons License or equivalent. Shah H and Golder P. ADaPPT: Enterprise Architecture Thinking for Information Systems Development. International Journal of Computer Science Issues, *8,1 (2011), ISSN (Online): 1694-0814.*

timely and understood in order to facilitate appropriate analysis and to appreciate the relevant impacts of decisions made. Organisations are continually faced with the challenge of their IT delivering the business value demanded and responding speedily to the changing business needs.

8.2 ENTERPRISE ARCHITECTURAL THINKING

An EA approach can help to provide a vehicle for organisational communication. Improving communication and discussion between business and IT staff enabling a shared understanding of the business and its supporting infrastructure that can facilitate improved decision making and more effective deployment of change. The approach provides a basis for standardisation and agreed notations and representations, processes and information become more transparent. Project costs can become more stable and better predicted, the time taken to bring about change either by enhancing current services or by offering new ones can be reduced.

Clearly identifying the key components through an enterprise architecture approach of business processes, information, technology applications and organisation and how these relate to each other facilitates focussing on the appropriate component as required in a particular situation. EA can be used to manage complexity and describe the interdependencies in a usable manner.

EA can facilitate a better return on an organisations investment by providing a means to identify cost saving opportunities, gaps and inconsistencies as well as facilitating the installed systems and applications being exploited. An enterprise architecture approach leads to improved scoping and coordination of programmes and projects.

8.3 THE CHALLENGE OF CHANGE

It is commonplace to identify the forces of change to which modern businesses are exposed. It is relevant to discriminate between forces for change that affect the business being carried out and those which affect only the way the business is delivered. A new computer system leaves organisational models unchanged but may change the data models and applications used to support them. A change in market or a merger will change the organisational model itself. The pressures for change on the organisation are such that a process of continual evolution even revolution is affecting all organisations. This means that the construction of an enterprise architecture is not a single event generating a static description of a the organisation which thereafter impedes the process of change. On the contrary the continual evolution of the enterprise architecture is a process in parallel with the evolution of the business strategy. The question should be asked how do we architect the business to meet its evolving strategic needs and the answer should lie in the continual evolution of the architecture. The architecture is the interface between the strategic, what the enterprise wants to do, and the operational, what it does.

Strategic change in the organisation can lead to evolutionary changes in the enterprise architecture but may require more radical change. For example the merging of two organisations may require the integration of their existing enterprise architectures into a new common EA. This is a similar problem to the evolving enterprise architecture one but is likely to require more substantive change. For example we may not be able to assume that the concept 'Customer' is exactly the same in the two merging organisations so may need to examine this at some detail in order to achieve successful integration. However in an organisation that is evolving from concrete to virtual trading, the concept of customer may be undergoing equally significant change and the significance of this change may be overlooked in the assumption that it is evolutionary.

An organisation has a business strategy at a particular time; corresponding to this strategy it has (or is in the process of developing) the corresponding enterprise architecture for delivering this strategy. That existing enterprise architecture describes and specifies a number of business processes, data objects and applications which 'operationalises' the architecture. Next the business introduces a new strategy, corresponding to this we have desired enterprise architecture and its corresponding business processes, data objects and applications. The practical problem becomes how do we migrate from one EA to the next? Moreover we would want to know the series of architectures (or roadmap) that would take us through the required transitional architectures. How do we identify the changes necessary in business processes, data objects, and applications required and how do we manage the transitions.

The following examples serve to highlight what is needed in the management of organisational change through evolving enterprise architectures.

- We need to be able to examine the ontology of concepts—what is a customer?
- We need to be able to identify the dynamics of the elements—how does a customer come into existence, what determines the life of a customer, how is it terminated?
- We need to be able to identify the agents responsible—who authorises the creation of a customer, who determines when a customer is no longer?
- We need to be able to specify the business rules related to the behaviour of customers and agents.

Existing methodologies and tools do not help use with these problems they are mainly focused on the storage and retrieval of data, and the specification of data manipulation processes. An enterprise architectural approach can facilitate this thinking.

Enterprise architecture has been widely adopted as a means to cope with the ever-increasing complexity of organizations and to ensure that the technical resources are appropriately employed and optimized [1]. EA is the fundamental organization of the system, embodied in its elements, their relationships to each other and to the environment and the principles guiding its design and evolution [2], [3]. Enterprise architecture is described as organizing logic for business processes and IT infrastructure, reflecting the integration and standardization requirements of the company's operating model in order to achieve business agility and profitable growth [4]. Currently, there exist a number of professional societies and organizations that are working on the definition and the management of enterprise architecture such as The Open Group, Microsoft, and IBM. Indeed, EA represents much more than IT architecture. It is an integrated and holistic vision of how the business processes across the enterprise, people, information, applications and technologies align to facilitate strategic objectives. EA frameworks identify the scope of EA and decompose various elements of the architecture onto structured layers/levels and elements. Several EA frameworks have been adopted for operational use in many private and governmental organizations.

EA emerged as an idea in 1980 and is embodied in the early EA framework developed by Zachman (1987) [5]. EA has re-emerged as a means to cope with the ever-increasing complexity of organizations. This re-emergence is closely related to the evolution of new business trends and to the evolution of IT, particularly to the advances in Internet technologies. These business trends comprise globalization, mergers and acquisitions, e-commerce, as well as the increasing importance of customer relationship management (CRM) and supply chain management. IT trends, on the other hand, comprise the advances in Internet technologies, hardware platform, application servers, and workflow servers. Due to the increasing importance of EA, certification opportunities in EA are being offered by several companies such as The Open Group and IBM in order to standardize an open method for IT architecture to solve business problems.

An EA approach is beneficial in aligning business and IT resources and in conforming to fundamental principles and common methodologies that govern the entire life cycle of the IS development process. In that sense, architectural frameworks are considered to be a convenient way to support such methodologies, and to separate roles that facilitate and implement these methodologies as needed. Still, there are many organizational and technical EA challenges.

8.4 ENTERPRISE ARCHITECTURE FRAMEWORK

EA frameworks identify the scope of the enterprise architecture and decompose various elements of the architecture onto structured levels and elements [6]. More formally, EA frameworks describe a method for designing IS in terms of a set of building blocks and how these blocks fit together. Several EA frameworks such as ARIS [7] and DODAF [3] have been adopted for operational use in many organizations. For example, the Federal EA [8], has been adopted by the US government as a business-driven framework in order to optimize some strategic areas. These areas include budget allocation, information sharing, performance measurement, and component based architecture. More specifically, EA frameworks contain a list of recommended standards and compliant products that are used to implement the building blocks for an IS. EA frameworks are useful in terms of simplifying architecture development and ensuring complete coverage of the designed solutions through a common terminology. In that sense, these frameworks are language independent by providing generic concepts and common terminology through which different EA stakeholders can communicate without making any assumptions about each others' language. Pragmatically, EA frameworks play a dual role. Firstly, they serve as implementation tools; secondly, they can serve as organisational planning tools.

8.5 THE ADAPPT APPROACH

ADaPPT was developed in work with organisations using the AL-TAR (Achieving Learning Through Action Research) methodology [1].

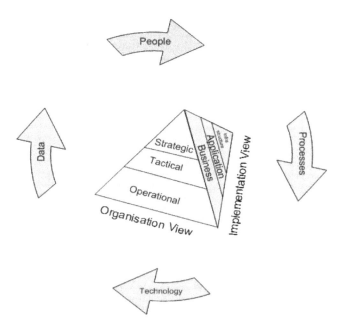

FIGURE 1: ADaPPT Approach.

ADaPPT has four domains (elements) (We call these elements because just as the ancients believed that every thing was composed of the four basic elements, Fire, Water, Air, Earth, so ADaPPT believes that every business activity combines the basic elements of, People, Process, Data, Technology): people, processes, technology and data.

8.5.1 ADAPPT: ALIGNING DATA, PEOPLE, PROCESSES, AND TECHNOLOGY

There are two main views in ADaPPT: a organisational view and an implementation view. The model recognises that everything the enterprise does involves people, processes, technology and data and that these need to be aligned.

"A process driven by people consumes resources (technology) and generates data."

In the ADaPPT framework the term people represents not only individual people but groups within the organisation, departments, sections etc. and also roles such as marketing manager etc. In as much as they can initiate actions and be responsible for the processes of the organisation. The term agent is also used.

In ADaPPT process means all activity and actions within the organisation. This includes high level business processes—marketing, production etc.; middle level activities—launching new products etc; operational activities—checking an invoice—etc.

In ADaPPT technology includes all services, material and equipment used by the organisation. This includes computer and IT hardware and software, raw materials and processed product.

In ADaPPT data means all information both static and dynamic within the organisation such as management targets, performance measures and operational data: customer details etc.

The organisational view in ADaPPT recognises that the nature the enterprise varies throughout the organisational hierarchy:

- Strategic
- Tactical
- Operational

As we climb the hierarchy processes become less well defined, soft data becomes more important, the scope of responsibility becomes wider. ADaPPT does not attempt to be a complete framework for all enterprise needs it is focused on the business/technology issues.

The implementation view recognises that there is a spectrum from business through application to infrastructure. This can apply equally to each domain. For example data can be viewed at: the business level expressed as E-R and other data models; at the application level expressed as data structure diagrams; at the infrastructure level expressed through allocation of data on storage devices. If we consider the interactions between views we will see that the implementation view and the business views are independent of each other. Thus an operational level business problem will need to be addressed at the business level as a specification of the business problem, at the application level as design of the solution to the problem and at the infrastructure level by provision of software and other resources to implement the solution.

The ADaPPT framework thus has four domains each of which can be described with a three by three matrix of views.

"This is because enterprise planning is a complex process involving many thousands of elements. The ADaPPT approach aims to organise and simplify the process of thinking about and managing these thousands of elements."

8.5.2 ADAPPT AS IMPLEMENTATION TOOL

ADaPPT in common with other EA frameworks provides a comprehensive representation of IS in terms of its building blocks. In this context, ADaPPT relates the necessary IS aspects/dimensions such as business processes, data, and organization units to different perspectives at certain levels of abstraction. These perspectives rely mainly on the difference in EA stakeholders' views of the architecture that span different level of details. EA frameworks, as components specification tools, encompass the documentation of the architectural layers, architectural domains, architectural models, and architectural artefacts.

Typically, EA frameworks such as ADaPPT are decomposed to three architectural layers, which are business layer, application layer, and technology infrastructure layer [9]. The business layer describes the business entities such as business processes and relevant business information, and how these entities interact with each other to achieve enterprise wide objectives. The application layer determines the data elements and the software applications that support the business layer. The technology infrastructure layer comprises the hardware platforms and the communication infrastructure that supports the applications. Such layers are naturally characterized by information aspects, behavioural aspects, and structural aspects. As organizations consist of several units, the structural aspects determine the static decomposition of these units to several sub-units. The behavioural aspects show behaviour manifested in the sequence of activities and business processes performed to produce the needed services. These units exchange information in order to carry out business tasks. Each layer is naturally composed of several domains that reflect the information, behavioural, and structural aspects of the organizations. These domains specify the architectural aspects such as process architecture, product architecture, information architecture, technical architecture, and application architecture. Indeed, these domains are the means to separate the architectural concerns and reflect the view of different EA stakeholders to the architecture. For example, the process domain, which is a part of the business layer, describes business processes or business functions that offer the products or services to an organization. These architectural domains are typically described and documented by different architectural models such as business process models, value chain diagrams, and organization charts. Architectural models serve as a basis for documenting the different architectures by annotating the artefacts and their inter-relationship that are needed to model an organization from different perspectives. Architectural artefacts represent the necessary constructs and architectural elements such as data, business processes, resources, and events that represent the real world objects needed to design distinct model types.

8.5.3 ADAPPT AS IMPLEMENTATION TOOL

ADaPPT in common with other some other EA frameworks provide a holistic view of EA through the hierarchical layering, which implies the alignment between, business, application, and technology infrastructure layer. As such, business decisions and architecture planning can be made in the context of whole instead of standalone parts. In other words, EA frameworks such as ADaPPT make use of the abstractions in order to simplify and isolate simple IS aspects/dimensions without losing sense of the complexity of the enterprise as a whole. As an organisation planning tools, ADaPPT entail baseline architecture, future architecture, architectural roadmaps, and transition plans. Baseline architecture, which is also known as 'as-is' view, encompasses the documentation of different layers and the existing components (models, diagrams, documents etc). This architecture serves as a baseline for identifying the relationships between different components and the gaps that should be filled for better organizational performance. Target architecture, which is also referred to as the 'to be' view, specify the new EA components and the strategic initiatives that should be carried out for the sake of bridging the gaps and ensuring competitive advantage. This architecture should also identify the IT resources and technological infrastructure that are needed for supporting the new EA components in order to integrate the organization structure, business processes, data, and technical resources. Architectural roadmaps represent the intermediary EA alternatives of the baseline architecture generated in the process mitigating the risks and analyzing the existing gaps in order to shift to the target architecture. These roadmaps annotate the architectural milestones performed prior to reaching the target architecture. EA transition plans are merely specifications of an 'as-is' and 'to-be' view in terms of managing the feasibility of architectural transition such as risk assessment, gap analysis, and the supporting resources of the transition. More specifically, transition plans document the activities that need to be undertaken to shift from the baseline architecture to the target architecture. Such plans are means to determine the desired future state of the enterprise wide goals, business processes, technical resources, organization units, and data.

AdaPPT Domains	Data	Zachman Function	Network	People	Time	Motivation
People			X	X	X	X
Process		X	X		X	
Technology			X		X	
Data	X		X		X	

FIGURE 2: ADaPPT and Zachman Framework.

8.5.4 ADAPPT AND OTHER EA FRAMEWORKS AND TOOLS

A range of tools can be used to model the architecture appropriate to the different views. Where appropriate familiar tools are used across several views so we do not need 36 different models as in the Zachman [5] approach eg E-R modelling is used for the Data Domain and UML can be used in the Process Domain Business and Application Views. Figure 2 illustrates a mapping between the Zachman approach and ADaPPT.

One of the leading EA toolset is ARIS [7]. Whilst a toolset may support many frameworks it will also have an implicit ontology. ARIS is a complex tool with an underlying Process Model. ARIS manages complexity with four Views: Data View, Organization View, Function view, Product Service View. ARIS supports a detailed process oriented view of the organisation. However the basic units of ARIS fit easily within the ADaPPT framework. There is no inconsistency in using ADaPPT as an EA Framework and ARIS as the toolset to support the management and operation or the Enterprise's Architecture Repository. The four ARIS views are apparently consistent with the ADaPPT framework. However it is worth examining some of the lower level ARIS concepts to see if this apparent alignment in high level concepts is reflected in the detail (see Figure 3).

In ADaPPT it is possible to use conventional diagramming tools such as MS Visio or complex EA diagrammers or even full blown modellers

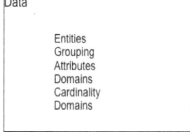

FIGURE 3: ARIS concepts within ADaPPT Domain.

such as IDS Sheer. It is the way the different elements combine which creates business value. The main relationships in ADaPPT (see figure 4) are:

- People: Initiate processes, Use data, Specify technology
- Processes: Run on technology, Use technology, Generate data
- Data: Stored on technology

These can be represented in the various EA tools. These main relationships are important. Representing, exploring and planning related to these elements is facilitated by, recognising and taking account of the main relationships and their content from organisational and implementation perspectives as appropriate for the particular context under consideration.

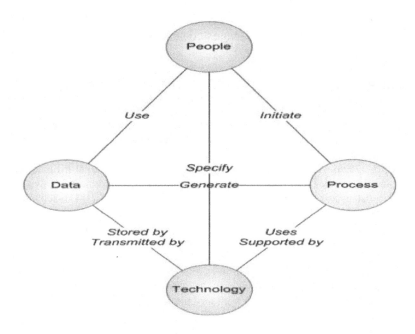

FIGURE 4: The main relationships in ADaPPT.

8.6 CONCLUSIONS

Our ongoing work is applying ADaPPT based enterprise architecture thinking for information systems development in public and private sector organisations. It provides a strong foundation for understanding the strategic, managerial and operational issues in aligning people, processes, data and technology and also in developing strategic, managerial and operational approaches while ensuring the alignment of people, processes, data and technology in an IS context. It is being used as the basis for understanding how knowledge management can be improved and new technologies exploited by organisations. Providing a basis for enabling the conceptualising of holistic, integrated and detailed consideration as appropriate to the development stage and stakeholder perspective.

REFERENCES

1. Author, Eardley WA and Wood-Harper AT (2007) ALTAR: Achieving Learning Through Action Research, European Journal of Information Systems, Vol 16 No 6, Pp. 761-770.
2. IEEE (2006), "IEEE Standards Association, IEEE Std 1471-2000 IEEE Recommended Practice for Architectural Description of Software-Intensive Systems, http://standards.ieee.org/reading/ieee/std_public/description/se/1471-2000_desc. html," 2006.
3. DOD (2006), "DOD Architecture Framework, Systems and Software Consortium, http://www.software.org/pub /architecture/dodaf.asp," 2006.
4. J. W. Ross, P. Weill, and D. C. Robertson (2006), Enterprise Architecture as Strategy: Creating a Foundation for Business Execution. Boston, Massachausettes: Harvard Business School Press Book.
5. J. A. Zachman (1987) "A Framework for Information Systems Architecture," IBM Systems Journal, Vol. 26, No. 3.
6. Author and El Kourdi M (2007), Enterprise Architecture Frameworks, IEEE IT Professional, Vol 9, No 5.
7. A.-W. Scheer (1999), Business Process Engineering: Reference Models for Industrial Enterprises, 2nd ed. berlin: Springer.
8. FEA (2003) CIO, " Practical Guide to Federal Enterprise Architecture," Enterprise Architecture Program Management Office, http://www.feapmo.gov.
9. H. Jonkers R. v. Buuren, F. Arbab, F. d. Boer, M. Bonsangue, H. Bosma, H. t. Doest, L. Groenewegen, J. G. Scholten, S. J. B. A. Hoppenbrouwers, M. E. Iacob, W. Janssen, M. M. Lankhorst, D. v. Leeuwen, H. A. Proper, A. Stam, L. v. d. Torre, and G. E. V. v. Zanten, (2003), "Towards a Language for Coherent Enterprise Architecture Descriptions," in 7th IEEE International Enterprise Distributed Object Computing Conference (EDOC 2003). Brisbane, Australia, 2003.

CHAPTER 9

A NEW METHOD OF PERFORMANCE EVALUATION FOR ENTERPRISE ARCHITECTURE USING STEREOTYPES

SAMANEH KHAMSEH, FARES SAYYADI, and MOHAMMAD HOSEIN YEKTAIE

9.1 INTRODUCTION

These days, most organizations use the recent advances of information technology for making strategic decisions. Many organizations have complex infrastructure with improper architecture, low efficiency, flexibility and speed to transfer the information. Enterprise architecture is a set of representations or models described in connection with a description of an organization, and the primary objective is to manage and use necessary items. Architecture includes a large number of documents, which describe all parts of organizations (Deft, 2000). The problem with this description is on how they all be noted and be used. Therefore, to create order and organization of the enterprise architecture description, a framework needs to be applied. There are some shortcomings on modeling notation to cover all C4ISR products and enterprise is one of the most important challenges facing the C4ISR architecture framework. The necessity of such a modeling notation where the use of a variety of symbols and language modeling for cover crops, causing confusion and inconsistency and architecture

This chapter was originally published under the Creative Commons License or equivalent. Khamseha S, Sayyadib F, and Yektaiec MH. A New Method for Performance Evaluation of Enterprise Architecture using Streotypes. Management Science Letters _3 (2013), DOI: 10.5267/j.msl.2013.10.012._

work is hard and complicated. The task of mapping products is normally accomplished using Enterprise Architecture Framework C4ISR, which is established by unified modeling language (UML) as an object-oriented approach (Sowa & Zachman, 1992). However UML is unable to express the needs of vague and indeterminate instances because the needs of users for information systems is based on the formation, Therefore, access to some of the system's requirements is a challenging task. In order to overcome these shortcomings, the Fuzzy-UML is implemented (Zadeh, 1983; Bostan- Korpeoglu & Yazici, 2006). In any enterprise architecture, the process name, process enterprise architecture and the three-phase development strategy, planning, architecture and implementation of the architecture need to be clearly specified. According to Lindsay et al. (2003) analysis of enterprise architecture planning phase can be accomplished to evaluate the behavior and the performance evaluation. Since most software systems are unable to handle all necessary needs in evaluating enterprise architecture, to evaluate these kinds of systems they must first create an executable model. In this paper, to do this, a fuzzy colored Petri nets are proposed. There are many ways to create an executable model of the software and its evaluation. However, the primary objective of this paper is based on an execution model using stereotypes fuzzy UML diagrams. To evaluate the behavior of enterprise architecture, the fuzzy UML diagrams are used. However, in this paper, to create executable models of software systems, the existing stereotypes in the use case diagram, sequence and deployment are applied. One of the most important parameters implemented to evaluate a software system is associated with reliability of a system and a system with high level of reliability ensures long-term performance. In other words, reliability is defined as the probability that a system will work, properly under pre-defined circumstances. In this paper, by using an executable model of the stereotypes created by the three fuzzy UML diagram above, we improve reliability of the system (Kaisler et al., 2005).

The remaining paper is organized as follows: The second part of the paper is devoted in enterprise architecture description. In the third part of the paper, we propose a method to evaluate the reliability of enterprise architecture. The fourth section of the paper evaluates the results and findings of the proposed algorithm implemented on a case study and the results are addressed. Finally, in the fifth section of the paper, conclusions and future work are described.

9.2 LITERATURE REVIEW

There are various techniques to evaluate the architecture and the aim of this paper is to investigate the effect of architectural styles, which is essential that the techniques based on the models and simulations used to evaluate the architecture. To use this technique, it is necessary to first develop a model to analyze and to evaluate architectures. Both models are mainly for work done: First, the architecture products are designed, under certain modeling language, and products and architecture, are commonly created by standard modeling language like UML. Next, using these products, we create a model for evaluating the architecture. For the ARCHIMATE model, architecture is evaluated in terms of performance. With the normal practices of the organizational model and its parameters, this model is suitable for quantitative analysis of architecture. For the proposed model, we can quantitatively analyze the architecture, the language, the characters and relationships are added to the context in which certain quantities can be determined. There is a method developed by Levis (Shin et al., 2003), which creates an executable model and the main advantage of this method is that Architecture Modeling Language UML is implemented in this model, which is a popular technique. In this way, the enterprise architecture implemented C4ISR architecture products are produced in this context and colored Petri nets are applicable for this methodology (Wagenhals et al., 2003). OSAN is one of the powerful language modeling for evaluating the architecture, especially the architecture performance evaluation. Bai et al. (2008) applied a model of the architecture caused. The advantage of this modeling language is that the model fully supports object-oriented programming. In recent years, much attention is devoted on evaluating enterprise architecture. JavadPour and Shams (2009) used C4ISR as an architecture framework with regard to the fixed frame as architectural structure, to evaluate the performance of the software architecture with different behaviors (different styles) with Network Colored Petri Nets. Mozaffari et al. (2011) performed an investigation on enterprise architecture to analyze and to evaluate enterprise architecture knowledge of software architecture design, to achieve appropriate architecture. Rezaei and shams (2009) presented a solution, which extracts the enterprise architecture federal enterprise architecture framework. They also defined the maturity level of the enterprise architecture that includes a detailed assessment of the existing architecture. Here is emphasize is on the uncertain nature of the system providing a technique for increasing the reliability of the system (Behbahaninejad et al., 2012).

9.3 THE PROPOSED METHOD

This article focuses on assessment of the performance of enterprise architecture using stereotypes UML. UML is a standard language for describing semi-formal enterprise architecture, which is easy to address functional needs and to address uncertainty we use fuzzy terms and develop F-UML (Ma, 2005; 2011b). Since UML is not a formal model, evaluation of software systems is not possible, directly and we need to have the access to the actual model. In this case, for the structural aspects of system, we use case diagram, sequence diagram and deployment diagram to show the behavior of the physical aspects of the system resources (Bernardi & Merseguer, 2007). The proposed solution using stereotypes performance in F-UML is suitable for modeling and evaluating the performance of enterprise architecture. Fig. 1 shows details of our implementation.

9.3.1 STEREOTYPES USED IN USE CASE

Generally, any user on the use case represents a sequence of requests in the system. This graph has the following stereotypes:

> 1- <<PAopenload>>: Used in cases where the request sequence is infinite that Tags <<PAoccurrence>> that is, the time between two consecutive requests shows.
> 2- <<PAcolsedload>>: Used in cases where the request sequence is limited and contains the following tag:

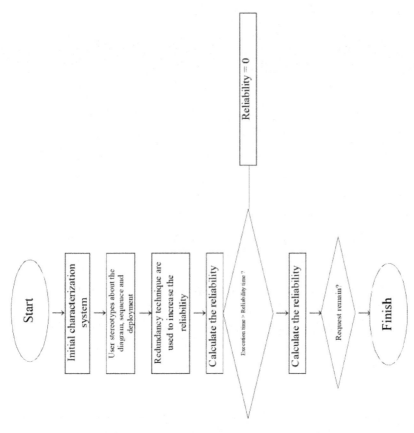

FIGURE 1: The flowchart of the proposed algorithm

1-2- PApopulation: Shows the total number of requests in the system.
2-2- PAextDelay: Full time interval between a request and subsequent interaction with the system show.

For the proposed algorithm, the use case diagram and label the stereotype << PAclosedload >> PApopulation, PAextDelay, are used.

9.3.2 STEREOTYPES USED IN SEQUENCE DIAGRAMS

In this diagram, all existing interactions in the system are displayed. To add efficiency considerations on the sequence diagram, the stereotype << PAstep >> is used. These stereotypes include all of the following labels on system reliability assessment, which are used:

1- Label size: It specifies the size of the message.
2- Label demand: Rate will apply to the supply of services.
3- Label PAhost: The name is a reference to the requested resource.
4- PAprob: Indicates the likelihood of the message.
5- PArep: Indicates the message is repeated.

9.3.3 STEREOTYPES USED IN DEPLOYMENT DIAGRAMS

Deployment diagram explains how to get a picture of the physical system resources. In this diagram, for additional performance information, the stereotype << PAhost >> uses labels that include the following:

1- Label PArate: Shows the processing rate.
2- Label Schdpolicy: Policy schedule shows.

In our proposed algorithm, to assess the reliability of the system, we will also use these two tags.

To enhance system reliability, the proposed algorithm uses the redundancy technique. To this end, each message is assigned to more than one component for parallel execution. Therefore, if one component fails dur-

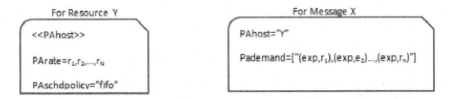

FIGURE 2: The proposed stereotypes with tags

ing execution, by replacing the components of the same source, we will prevent the failure process and, thereby, we will increase system's reliability.

To add redundancy to the system, only two stereotypes, PArate, PAdemand, will be manipulated. Thus, if we have n elements from a source system, then the deployment diagram, n will have to PArate and graphs can be arranged from 1 to n to do pAdemand. The following figure illustrates this better than the words:

We need to calculate reliability:

If the CPU processing rate is rpp (PArate), service request rate from the source is rpd (PAdemand) name, the first step is to calculate the service rate as follows:

$$SR[i, X] = \frac{rpp[i]}{rpd[i]} \tag{1}$$

Order of SR[I,x], the service rate of component i in the X source.

The service rate for each component of the resources at run time to get the message size z with y (size) of component i, going to X source is calculated as follows,

$$T[i, X, Y] = \frac{y}{SR[i, X]} \tag{2}$$

To the T[i, x, y], y is the size of the message by the time ith component of x as a source.

Since the proposed algorithm increases the reliability of the system, using the redundancy technique, the simulations is carried out and each message may be processed by several sources. Let $T_{x,r}$ represents the minimum time of the message x, therefore we have,

$$T_{x,r} = \min[T(i, X, y)] \qquad 1 \leq i \leq n \tag{3}$$

For the execution of a task, all messages must be executed, so the time to run messages, to run the greatest time of task execution time will be considered. However, if the messages are dependent on each other, the total running time of all times, as summarized in the task. Namely,

$$T = \begin{cases} \max \ (T_{x,r}[i] & 1 \leq i \leq m & \text{if no correlation} \\ \displaystyle\sum_{i=1}^{m} T_{x,r}[i] & 1 \leq i \leq m & \text{else} \end{cases} \tag{4}$$

A completion of a task cannot be longer than logged and system reliability can be calculated as follows:

$$R_{T^*} = \sum_{i=1}^{I} P_i \cdot l(T_i < T^*) \tag{5}$$

We represent the entire task entered into the system. Let T* be the same as $T_{reliability}$. T_i computation time to run the task with the number I as follows,

L(true)=1, L(false)=0 (6)

Let P_i be the number of running task, which is calculated as follows,

$$P_i = \prod_{j=1}^{m} q_j \tag{7}$$

where q_j denotes the probability of the message with the number j of task i, which will be calculated as follows:

$$q_j = p(e)^{bp} * p(d) \tag{8}$$

where P(e) is the possibility that e constituents working during processes, P(d) is the probability that communications channels is likely to be healthy, bp is the number of components and e is busy status.

In our proposed algorithm, if a message is performed by multiple sources, to calculate the reliability we have:

$$Reliability = 1 - \prod_{j=1}^{n}(1 - R_j) \tag{9}$$

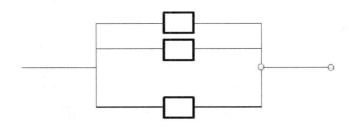

FIGURE 3: Parallel system with resource

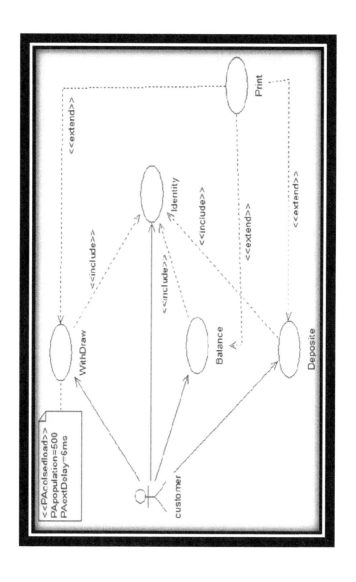

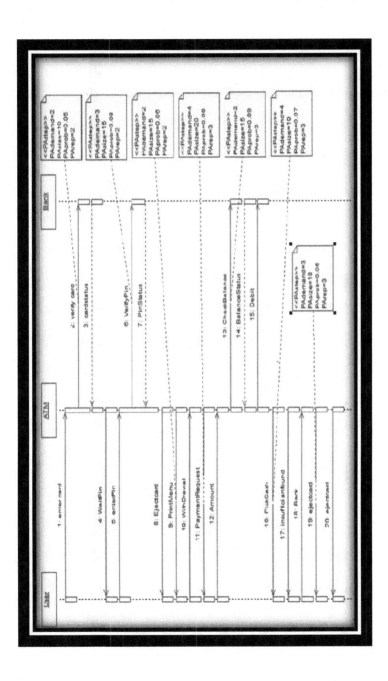

FIGURE 5: Sequence Diagram for ATM

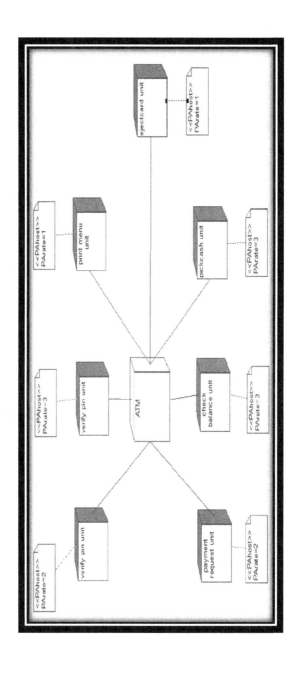

FIGURE 6: Deployment Diagram for ATM

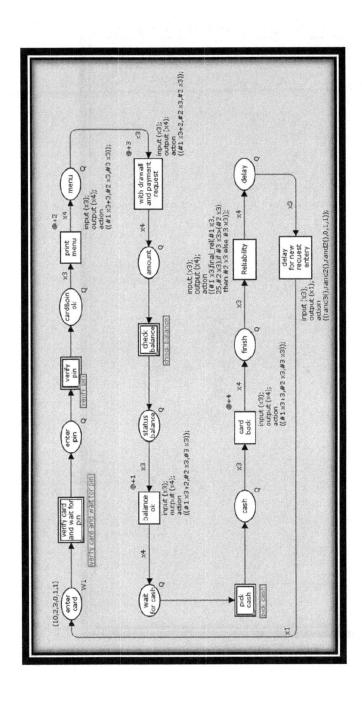

FIGURE 7: The high-level model of ATM

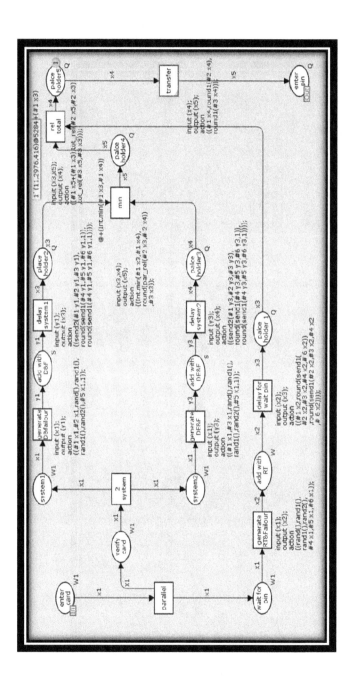

FIGURE 8: Subpage OF verify card and wait for pin

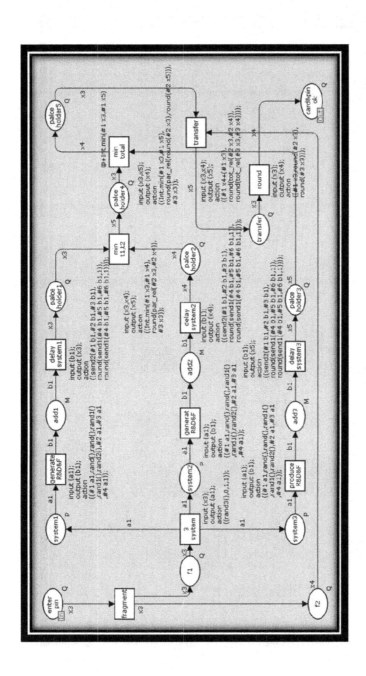

FIGURE 9: Subpage for wait pin

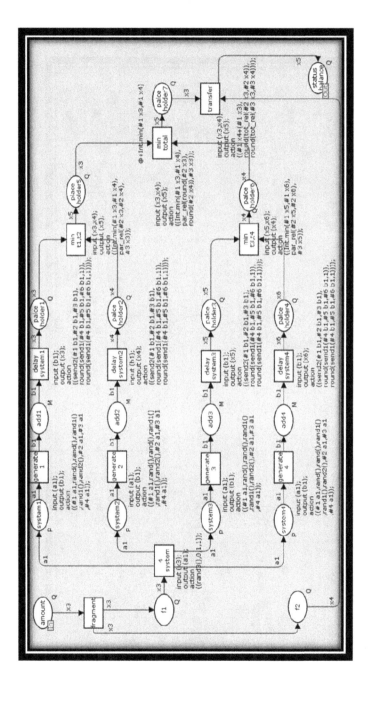

FIGURE 10: Subpage for check balance

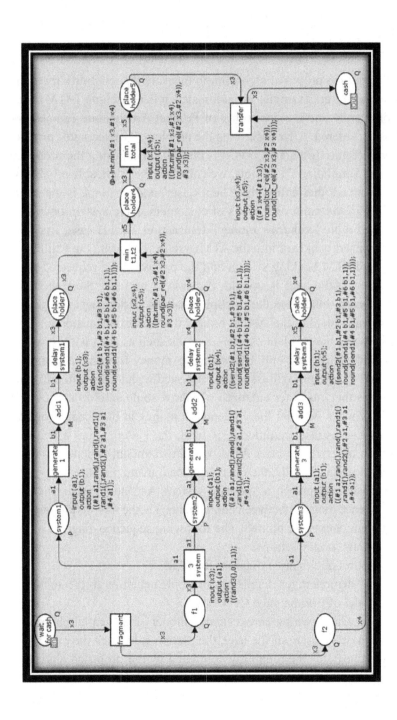

FIGURE 11: Subpage for Pick cash

9.6 CASE STUDY

We now apply the proposed algorithm to measure the reliability for some case studies. As an example, an automated teller machine (ATM), due to its complexity and the possibility of Product Architecture Framework C4ISR is considered. After describing the problem, the executable product model architecture (diagrams UML) is created. Then, using the proposed method and simulation CPN tools, we calculate the applicable reliability. An ATM with another entity, the Client (User) and the Bank is the interaction. One of the main operations of customers in the system operation is associated with "withdraw money" transaction. In this case, first, the user has to insert bankcard into the ATM system. The bank, the validity of the card, and ATM banking system and the password are requested from the customer. If the card is invalid, the system retrieves the customer's bank ATM card, otherwise the credibility of the customer is the key. If the password is invalid, the system returns the card to the customer. Otherwise, it shows the available system options such as cash withdraw and transfer funds. Customer-options "withdraw money from the account" is selected and the system will issue the requested withdrawal amount. This is a message that customer enters to withdraw some amount money and the system checks whether there is enough money in the account or not. When there is insufficient amount of fund, the system gives a warning to customer, otherwise, a bank ATM, withdraw amount is deducted from customer's account. Finally, the client application and the back end of the card, the customer can return the card system. As stated earlier, the pattern of sequence diagram between components, i.e. the interactions among different components are plotted. The following sequence diagram of use case "withdraw money" shows.

Fig. 6 shows deployment diagram for ATM.

Fig. 7-11 demonstrates details of our implementation to create an executable model of a home page CPNTOOLS

However, after creating an executable model of sequential phase diagram and phase diagram of the user, the system calculates the reliability. Fig. 12 demonstrates the reliability computed by our proposed algorithm and comparison the reliability with an existing method.

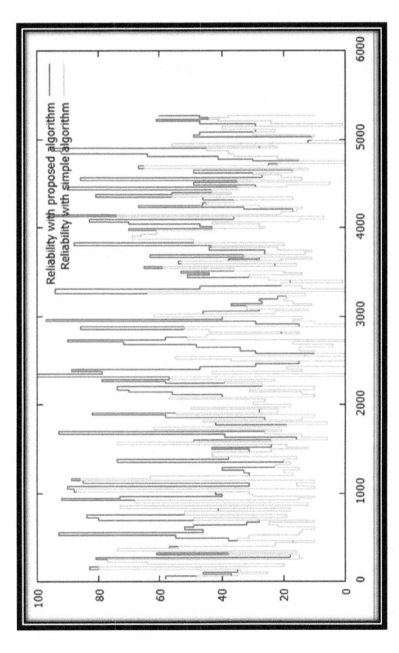

FIGURE 12: Compared proposed method with previous method

As we can observe from Fig. 12, using the redundancy in the system increases the reliability of the system, significantly.

9.7 CONCLUSION

In this paper, an algorithm was presented in the diagrams using stereotypes of F-UML and the reliability of the system was measured. We first evaluated the performance of the stereotypes and some diagrams including use case and sequence were used for the deployment using the art of FUML technique. The performance of the proposed algorithm was compared with other algorithms. The redundancy technique has been employed to increase the reliability of the system. After creating an executable model, their algorithm on a case study has been implemented. The results indicate that the proposed algorithm provides more reliable results compared with other previous algorithms.

REFERENCES

1. Bal, X.H. (2008). An application with UML object-based Petri Nets for C4ISR architecture simulation validation. proceeding of a seventh international conference on machine learning and cybernetics, Kunming.
2. Behbahaninejad, P., Harounabadi, A., & Mirabedini, S.J. (2012). Evaluating software architecture using fuzzy formal models. Management Science Letters, 2, 469–476.
3. Bernardi, S., & Merseguer, J. (2007). Performance Evaluation of UML Design with Stochastic Well- Formed Nets. The Journal of Systems and Software, 80, 1843–1865.
4. Bostan-Korpeoglu, B., & Yazici, A. (2006). A fuzzy Petri net model for intelligent database. Data & Knowledge Engineering, 8, 112-122.
5. Deft, R. (1983). Organisation theory and design. New York: West.
6. Javadpour, R., & Shams, F. (2009). Performance evaluation of electronic city architecture using colored Petri nets. The 2nd Conference on Electronic City, Tehran, May 2009.
7. Kaisler, S. H., Armour, F., & Valivullah, M. (2005). Enterprise architecting: Critical problems. Proceeding of the 38th Hawaii International Conference on system sciences, 8(8), 224.2.
8. Lindsay, A., Downs, D., & Lunn, K. (2003). Business processes—attempts to find a definition. Information and software technology, 45(15), 1015-1019.

9. Ma, Z. (2005). Fuzzy Information Modeling With the UML. Idea Group Publishing, 153-176.

10. Ma, Z.M., Zhang, F., & Yan, L. (2011a). Fuzzy information modeling in UML class diagram and relational database models. Applied Soft Computing, 11, 4236-4245.

11. Ma , Z.M. , Yan , L., & Zhang , F. (2011b). Modeling Fuzzy information in UML class diagrams and object oriented database models. Fuzzy Sets & Systems, 186, 26-46.

12. Rezaei, R., & Shams, F. (2009). Providing a comprehensive method for developing and evaluating enterprise architecture plan. The first Conference on Enterprise Architecture in Practice, Isfahan, Iran.

13. Shin, M. E., Levis, A. H., & Wagenhals, L. W. (2003). Transformation of UML-based system model to design/CPN model for validating system behavior. In Proc. of the 6th Int. Conf. on the UML/Workshop on Compositional Verification of the UML Models.

14. Sowa, J. F., & Zachman, J. A. (1992). Extending and Formalizing the Framework for Information Systems Architecture. IBM Systems Journal, 31(3), 590-616.

15. Wagenhals, L. W., Haider, S., & Levis, A. H. (2003). Synthesizing executable models of object oriented architectures. Systems Engineering, 6(4), 266-300.

16. Zadeh, L.A. (1983). The role of fuzzy logic in the management of uncertainty in Expert System. Fuzzy Sets Systems, 11, 199-227.

CHAPTER 10

SIX SIGMA DRIVEN ENTERPRISE MODEL TRANSFORMATION

RAYMOND VELLA, SEKHAR CHATTOPADHYAY, and JOHN P. T. MO

10.1 INTRODUCTION

The volatility of current business environment requires companies to adapt to new processes and systems to satisfy customer requirements as well as remaining competitive (Rho et al, 2001). Business enterprise is inherently a complex entity. There are many risks involved in the changes that an enterprise needs to go through in order to transform itself to a more competitive form (Beasley et al, 2005). Typical risks include collaboration, confidentiality, intellectual property, transfer of goods, conflicts, opportunity loss, product liability and others. Inappropriate actions can be taken if the information is out of date (Kutsch & Hall, 2005) or the employee performance can be seriously affected (Lin & Wei, 2006). Kwon et al (2007) reported that those enterprises going through significant organisational change and downsizing of IT function was not simply reducing the workforce in the IT department. It also eliminated communication and information-processing conduits necessary for effective communication and coordination.

This chapter was originally published under the Creative Commons License or equivalent. Vella R, Chattopadhyay S, and Mo JPT. Six Sigma Driven Enterprise Model Transformation. International Journal of Engineering Business Management, *1,1 (2009), pp. 1-8. DOI: 10.5772/6787*

The uncertainty in business environment presents many research opportunities implementing engineering changes in enterprises. Rouse (2005a) found that value deficiencies and work processes defined the problem of Enterprise Transformation and that many fundamental changes addressed value from the perspective of the customer. Rouse also discussed how the problem solving and decision making ability of management could influence the outcomes of transformation, but did not elaborate on specific tools to guide the transformation. Yin and Shanley (2008) proposed a three dimensions model that could assist decision makers to merge or form alliances. Oberg et al (2007) presented the concept of "network pictures" as the modelling framework to illustrate and analyse changes in managerial sensemaking and networking activities following a enterprise change. They concluded that following a major enterprise transformation managers may need to adapt their previous network structure in a radical way.

The dynamics of enterprise change was analysed by Marino and Zabojnik (2006). In their analysis, if new firms can enter the market quickly, it is more likely that enterprise change is motivated by efficiency improvement as opposed to increased market power. Thus, there is less reason to challenge the change as it comes internally. However, many enterprises that have problems making changes generally suffer from human or organisational resistance (Buhman et al, 2005; Corbett, 2007). It is clear from these studies that issues on enterprise change should be dealt with in a systematic fashion, supported by a methodology that assists the whole of enterprise to transform.

Enterprise modelling is best used to analyse the business in both manufacturing and service sectors in terms of complexities and context. Enterprise reference architectures provide a structure to understand enterprise activities, for example, promote planning, reduce risk, implement new standard operating procedures and controls, rationalizing manufacturing facilities. Dewhurst et al (2002) identified five key design issues in constructing a generic enterprise model (GEM) when they attempted to "design" the enterprise model. Study of enterprise architecture requirements in the last decade has been focused on how enterprises can be designed and operated in relatively static, authoritative environment (Molina & Medina, 2003; Mo et al, 2006). These enterprise engineering researches shared a common starting point, viz, stepwise, multi-dimensional enterprise modelling and design methodologies have been applied. The rationale to use enterprise engineering methodologies to guide these developments is to minimize the impact of uncertainty to enterprise performances as well as other associated processes (Ortiz et al, 1999).

There are limitations in the present enterprise modelling methodologies when applying them to modelling enterprise transformation. Current enterprise architectures are described tacitly with the assumption that the present state of enterprise does not change during the life cycle of the "enterprise engineering project" (Chen et al, 2008). In recent years, six sigma methodology (Jochem, 2006) has been embraced by many large and small corporations. The process focus of six sigma can benefit an enterprise by providing the means to develop a road map and initiate the required enterprise transformation that could become culturally more ingrained within the organisation. This paper examines the traditional approach of enterprise transformation by enterprise modelling design and explores how six sigma methodology can be used to facilitate a systematic enterprise modelling process providing a culturally embedded framework for enterprise transformation. We then propose a unified methodology incorporating six sigma in enterprise transformation and illustrate our proposed methodology by a case study.

10.2 ENTERPRISE TRANSFORMATION BY ARCHITECTURE DESIGN

To achieve enterprise transformation, architecture design approach uses the modelling formalism to create a baseline manufacturing enterprise model. The Generalised Enterprise Reference Architecture and Methodology (GERAM), which is an annex of ISO 15704, is a combined effort of an international task force (Williams et al, 1994). Based on a generic enterprise reference architecture, an enterprise model is captured as a business process engineering life cycle as shown in Figure 1.

Details of the physical, information and human structures are recorded by modelling formalisms and tools. For example, manufacturing processes are recorded as material flow charts. Management practices are captured as work flow diagrams. IT system architectures are captured in software

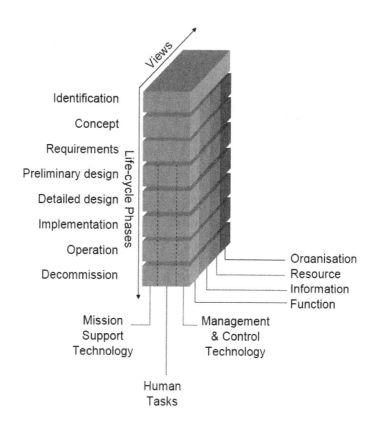

FIGURE 1: The Generic Enterprise Reference Architecture

engineering tools, and so on. The methodology helps enterprise engineers to encapsulate functions and processes within the enterprise. In a typical enterprise improvement process, the current state of the enterprise is captured. The problems in the current state are identified. The generic methodology helps the business process engineer to visualise "snapshots" that lead to the identification of trends and changes in the enterprise architecture (Figure 2). Over time, enterprise models are changed progressively. The outcome is a new enterprise model that describes the desired state of the enterprise at a particular time. The enterprise model outlines the elements of an enterprise engineering process that leads to suitable executable reference architecture for an organisation to deliver the responsiveness. They are useful for enterprises where internal functions are clearly defined and changes to parts of the system can be controlled at system component boundaries. After the enterprise model is developed, simulation technique can be used to evaluate the anticipated effect of enterprise design decisions to system performance (Love & Barton, 1996).

Traditional enterprise architectures are based on topdown approach. They emphasized on consistency throughout the organization and will involve all levels of employees. Normally, such a change is significant, since the organisation must have felt substantial pressure for a change that is more fundamental in nature. Changing the structure from current state to future state is often too long for dealing with the problem that the change intends to fix. The top-down approach also attracts inherent resistance to change from lower parts of the organisation. Furthermore, due to the fast changing business environment, the new model of the enterprise is a moving target. It takes a long time to progress from the current model to the newly designed enterprise architecture. When the changes are done half way through the transition, the enterprise designer is already under pressure to make further changes to the design under the new environment.

The enterprise architecture design approach focuses on designing the enterprise at different anticipated development stages using established enterprise design guiding principles (Uppington & Bernus, 2003). The success of the new (future) enterprise depends on the "correctness" of the enterprise vision and well managed implementation. Although simulation technique can assist in clarifying the effect of some design factors, there are many other aspects of the enterprise that can be simulated or foreseen. The transformation is therefore risky and nonresponsive to external environment. There is no systematic study of how an enterprise should be transformed to achieve a less risky but progressive path. A new approach is required to assist enterprises not only in defining their enterprise model, but also on re-engineering their processes and structures in a predictable way.

10.3 CHANGE PROCESSES IN SIX SIGMA

Everything in a business can be viewed as a process. Thus, an enterprise can be viewed as a collection of integrated processes interweaved with the four views of the enterprise in Figure 1. A manufacturing enterprise receives an order, schedules the production, builds the product, delivers the product and receives payment. A service enterprise receives a customer request, schedules a customer appointment, delivers the service and receives payment. For process improvement, six sigma has been well recognised as a powerful tool and as an imperative for achieving and sustaining operational and service excellence. While the original focus of six sigma was on manufacturing, today it has been widely accepted in both service and transactional processes (Jiju, 2004).

TABLE 1: DMAIC and DMADV processes

DMAIC		DMADV	
D	Define the project goals and customer (internal and external) CTS deliverables	D	Define the project goals and customer (internal and external) CTS deliverables
M	Measure the process CTS deliverables to determine current performance using verified measurements	M	Measure and determine customer needs and specifications
A	Analyze and determine the root cause(s) of the defects	A	Analyze and characterize the process options to meet the customer needs
I	Improve the process by eliminating defects	D	Design (detailed) and optimize the process to meet the customer needs
C	Control future process performance	V	Verify the design performance and ability to meet customer needs

Six sigma is a methodology for process improvement through reduced variability and the elimination of defects. Six sigma addresses system deficiencies using data to make decisions and formulating data driven solutions (Smith, 2001). The tool set is a collection of well known methods and techniques that is readily available. Intuition may be used to brainstorm, but all decisions are made using measurable data benchmarked against a set of Critical to Satisfaction (CTS) criteria, hence achieving measurable financial returns to the bottom line of an organisation.

Six sigma is structured as a sequence of processes. DMAIC and DMADV are two streams of processes dealing with specific changes in organisations. DMAIC methodology is used to improve existing products or processes that are not performing to target or not meeting customer expectation. DMADV is used when the process or product does not already

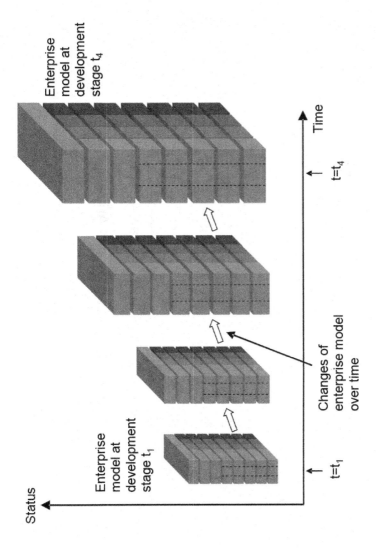

FIGURE 2: Enterprise models at different phases of enterprise development

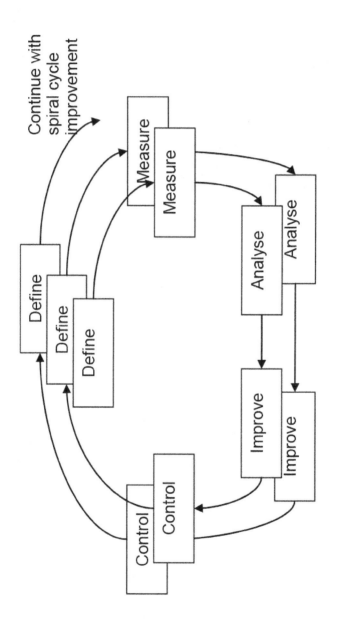

FIGURE 3: Spiral six sigma improvement cycle

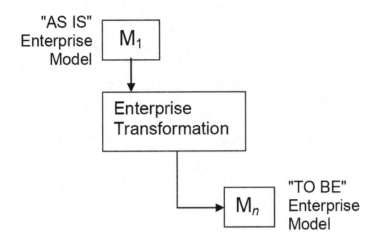

FIGURE 4: Enterprise model transformations

exist and one is required or when an existing process can not be optimised. Both streams consist of five steps as shown in Table 1. Six sigma processes are best represented as a cyclic spiral continuous improvement methodology (Figure 3).

Six sigma is a tool that can create a new process or improve existing processes. Its main scope of application has been on making incremental changes to particular segments of an enterprise such as cultural change, customer focus, process elements and statistical methods of improvement (Goh, 2002). Extension of this scope to an enterprise wide improvement is not common. Goel and Chen (2008) discussed business process re-engineering in the context of integrating a global enterprise using six sigma. They focussed on defining metrics for process analysis and refinement with the appropriate identification and analysis of tools to make the process transformation. However, the total picture of how the processes are linked in the organisation and the selection of which process should be transformed was not discussed.

The overall objective of enterprise transformation is to improve or redefine the inadequate business processes with appropriate tools to ensure the process requirements are satisfied and that relevant targets such as cost, delivery, productivity, and so on are met (McGinnis, 2007). Török (2008) emphasises the need to begin with the strategic business level so as to identify and confront the serious business challenges. These cascade to the operational level which identifies potential six sigma projects where each project then contributes to the strategic business requirement and resulting enterprise transformation. Rouse (2005b) argued that research in enterprise transformation should include transformation methods and tools, which should represent, manipulate, optimize, and portray input, work, state, output and value for the past, present and future of the enterprise. Six sigma needs a clear, transparent integrated definition and description of the processes of the enterprise for it to optimise and operate success-

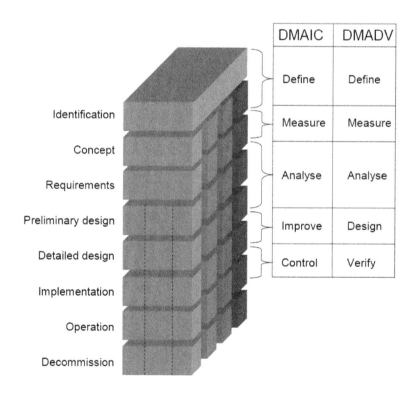

FIGURE 5: GERAM alignment with six sigma

fully. Since enterprise architecture design approach uses tools and methods that take a bird's eye view of an enterprise, we propose a methodology combining the two approaches to maximise global improvement outcomes during enterprise model transformation.

10.4 TRANSITION BETWEEN ENTERPRISE MODELS

Enterprise modelling provides a total enterprise view of processes, resources and technology. The level of integration, duplication or inefficiency in various processes may be derived through the development of the "As Is" enterprise model. Figure 4 shows a generic roadmap for enterprise model transformation. The existing business is represented with enterprise model M1 (the "AS IS" model). The forecast requirement is for an enterprise model as represented by enterprise model Mn (the "TO BE" model).

Enterprise modelling based transformation can be compared to six sigma methodology with the life cycle views of GERAM. Figure 5 shows how each of the five steps in six sigma would align with the phases of the enterprise lifecycle view.

In an "AS IS" enterprise model, the entire enterprise development is described as a snapshot of processes overlaying the four views. Irrespective of which phase a process has been developed, each process is explored to details as if it has been fully designed, implemented, operated and decommissioned. The enterprise is then regarded as fully described. If there is no change in identification and concept, which are often influenced by external parameters, the enterprise will progress through its lifecycle. Actual performance for each process may be benchmarked against targets thus highlighting processes with short fall performance. If the enterprise chooses, these processes can then be identified as candidates for six sigma projects, in which the processes are improved by focussing either on variability using the DMAIC process or redesigned using the DMADV process.

Six sigma has a narrow focus in its projects. Narrowness confines objectives in specific areas that maximise project support and ultimately aim at a successful conclusion. Transformation projects are selected by identifying what is critical to satisfaction (CTS) for the customer and what

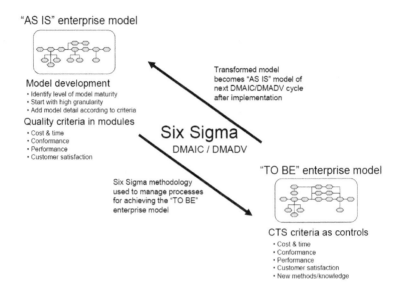

FIGURE 6: Transition plan with six sigma

areas of the business are deficient. Six sigma methodology contains tools to identify measurable deliverables required for customer satisfaction and identifies areas, processes or inputs that influence these deliverables (Smith & Fingar, 2003). Identifying and prioritizing these areas, processes or inputs against relevant criteria is one method that can be used to select six sigma projects that will lead to effective progressive enterprise transformation. The "AS IS" enterprise model can have varying levels of model maturity and granularity. Using six sigma, the business process engineer can focus on the relevant area and develop the necessary detail within the project scope. Thus, as shown in Figure 6, a six sigma project will take the "AS IS" enterprise model as the basis that represents the current real world condition and establish the relevant CTS criteria. This intuitive alignment between the two methodologies encourages the integration of six sigma methodology when generating new enterprise models. As the enterprise model is measured and analysed, six sigma methodology continues to refine CTS that will ultimately changes the concept and requirements of the enterprise with a context shift of who the customer is. Thus, as the context or view changes, this will enable the creation of, or improvement of, processes resulting in the "TO BE" enterprise model.

The added benefit of using six sigma is that a set of controls can be established for the process and the enterprise which should be included in the enterprise model. With dynamic market demands, the critical to quality characteristics of today would not necessarily be meaningful tomorrow. All CTSs should be critically examined at all times and refined as necessary. For this reason, a control plan developed from the last stage of the six sigma project should provide measures that indicate the performance and continued relevance of the processes within the transformed enterprise model. This can be an effective way of maintaining the enterprise model and relevance of the business to changing conditions.

10.5 PROGRESSIVE ENTERPRISE TRANSFORMATION PROCESS

The issue of enterprise transformation is the uncertainty of future enterprise models, in other words, Mn may not be fully described! This means the roadmap for transformation from M1 to Mn may not be defined until very late in the project causing significant disruption of the transformation process and turbulence in the enterprise. Change from M1 to Mn is normally significant due to the fact that an organisation will normally seek advice from a wide variety of resources to achieve the new design. This change is too abrupt and will impose significant organisational turmoil and reduce the ability of the enterprise to compete against other organisations.

Six sigma process has the advantage of self adjusting the system to suit the need of changes. The question is how to integrate six sigma tools as the method to make changes that are more than just continuous optimisation of a process. Linking six sigma with Enterprise Modelling to transform current state "AS IS" enterprise models to future state "TO BE" enterprise models with a clear road map is required.

If we follow the six sigma methodology, we will focus the transformation on process elements that have measurable CTS indicators relevant

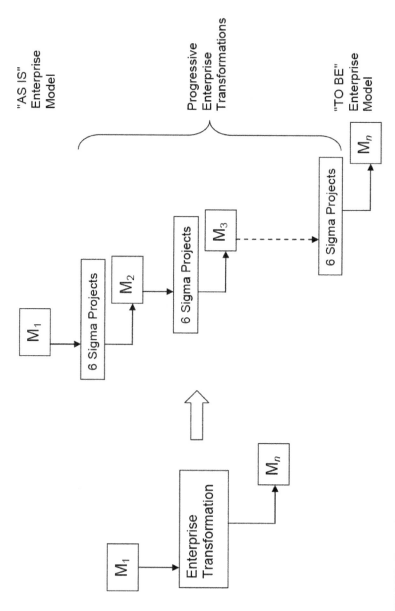

FIGURE 7: Enterprise model transformations with progressive six sigma driven changes

to internal business and external customer requirements. Therefore, as discussed in Figure 6, the CTSs form the foundation for decision making when redesigning or improving the process. The same indicators may form the controls to ensure continued performance of the new process or as indicators of when the new process is no longer relevant. Six sigma projects close the transformation loop by mapping CTS driven design requirements of the revised enterprise model.

Six sigma produces data driven solutions with relevant new and/or improved processes for the next "TO BE" enterprise model. It is therefore imperative that the initial "As Is" Enterprise Model contains accurate and relevant CTS information. It may be necessary to redefine the "AS IS" enterprise model to ensure that the CTS indicators and measurables are included and that the enterprise model does in fact represent the real life situation. This is in essence the first step (Define) of the six sigma project. The decision for any redesign efforts over traditional continuous improvement depends on a number of variables including risk, available technology & resources, cost, customer demand, time and complexity etc. This is part of the six sigma methodology and covered in the "Define" stage present in both DMAIC and DMADV based projects.

Instead of having an abrupt change between two major enterprise models, the new transition methodology divides the gap into many manageable steps (Figure 7). The intermediate "TO BE" enterprise models M_i ($1 < i < n$) in these steps are developed through six sigma projects. The changes are simple enough to be redefined. However, our emphasis is to make the change as smooth and manageable as possible to minimise impact. However, the future model M_n is defined to ensure a consistent approach or direction that governs each intermediate state to be processed by six sigma projects or a combinations of many grouped projects.

The road map developed for Enterprise Model transformation can be context related. The detail of each progressive enterprise model on the roadmap of enterprise transformation to M_n can be derived from six sigma projects with emphasis on the appropriate M_i design goals towards the ultimate state. This six sigma driven transformation becomes an iterative and progressive process of making changes leading to the larger enterprise model transformation in a similar manner to climbing Mount Everest. As the summit is ascended one step at a time, a new intermediate enterprise model may be developed until the summit can be reached in a final project. This progressive method of transformation is a logical approach for the enterprise model development that could become culturally more ingrained within the organisation already embracing six sigma for need based ongoing Enterprise Model Transformation (EMT).

In a more generic perspective, combining enterprise modelling with six sigma provides many benefits. Six sigma projects benefit from a big picture of the process lifecycle with the respective views or contexts. Enterprise modelling gains a tool set with methodology for transforming an "AS IS" enterprise model to the "TO BE" state. Within any stage of the lifecycle, the proposed alignment of the six sigma methodology to the generic enterprise architecture views remains the same. The changes from M_i to M_{i+1} is by methodological design to be incremental and manageable. The same tools and methods remain applicable during each of the steps used in transforming the different views of the enterprise model. The enterprise does not expect surprising disruption in this process and hence can progress smoothly through the enterprise model transformation.

10.6 CASE STUDY: ENGINEERING APPROVAL PROCESS CHANGE

We illustrate our proposed progressive enterprise model transformation approach by a case study at Ford Australia Pty Ltd. A process re-engineering project is presented here illustrating the use of six sigma as the tool to implement a needs based enterprise model transformation. The need for enterprise change was initiated by the long time required to approve engineering changes within the Product Development Organisation. From operational point of view, this change was driven by internal customer expectations. The case scenario is described below.

A new vehicle product development engineering service provided by Ford of Australia (FoA) to Ford India Private Limited (FIPL) required that an engineering change management process be extended to include approvals from FIPL. The initial implementation of the change manage-

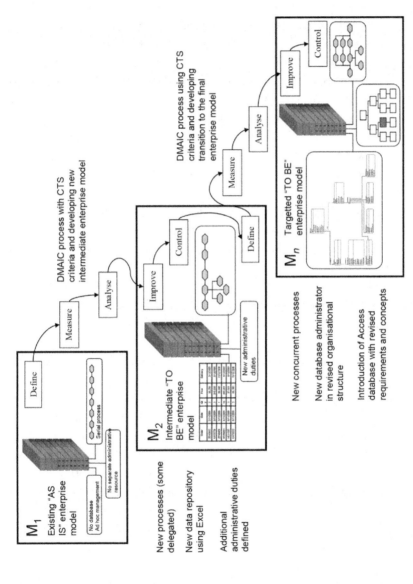

FIGURE 7: Engineering approval process enterprise model transformation

ment process was slow, cumbersome and did not meeting the needs of the business.

The change management activity was mapped including processes and organizational structure. A manifestation model of the present "AS IS" (M_1) state was produced for that section of the organization. The requirements critical to customer satisfaction were identified and stated as a problem definition—completing the first stage of the six sigma methodology (Define). The "AS IS" enterprise model was found to be a linear process with static work flow. It was not obvious what the process issues were or what departments produced the delays. Issues seemed to be regular in occurrence, but unique in nature. An "AS IS" model of the implementation phase was available, however, accuracy was uncertain and there were no clear measures or controls to identify process deficiencies or where changes may be required.

The desirable future enterprise model (M_n) would have many of the processes running in parallel and the ability to assure clear track and trace of work items in the system. There was also expectations of reduction of costs in the new model.

Following six sigma methodology, key process and resource elements identified in the updated manifestation Enterprise Model were used to develop suitable measurable metrics that would reflect the process performance. These metrics focused on critical to satisfaction outcomes that included activities contributing to the delay of engineering change approvals. The metrics allowed key elements of the process to be measured and areas of deficiency to be identified.

During the Measure phase of the six sigma methodology, it was clear that not all the metrics were readily available. Key performance measures could not be effectively measured with the current manifestation of the Enterprise. This was the driver for the first level of transformation, the use of a change management and approvals database. The database would automatically report the performance metrics against agreed departmental targets. This paved the way for the development of a road map that would define the next phase of Enterprise Model transformation with a new Operational Enterprise Model.

The initial minor transformation with the database provided metrics data that was analysed after one month using six sigma statistical methods.

It became clear how the new present model of this engineering activity was functioning and where the delays originated. The model identified a serial process flow. Performance metrics for each stage of the model were added presenting a clear view of what was happening in the enterprise. Certain model activities were not necessary. Some activities were independent and were realigned as parallel activities. A new "TO BE" model (M_2) was created with multiple parallel activities. Some new tools were introduced and data from different activities were integrated within the newly established change management database. In this case, poor change management performance was the deficiency. The transformation did not reduce variability. It changed the operational definition and introduced integrated tools to address values from the perspective of the customer. The Enterprise Model Transformation that has gone through Ford Australia is illustrated in Figure 8.

The six sigma methodology provided the framework to define the problem and apply the standard tools to identify metrics that would measure the existing "AS IS" performance. Interim progressive change was identified and actioned to help make the transition to the final model (M_2). The final Enterprise Model (M_2) would address areas requiring transformation to correct the operational deficiency. Appropriate metrics and data were used to make decisions on where to transform the business when the extent of the transformation was not clear. The metrics to measure and sustain the improvements were built into the new "TO BE" model.

10.7 CONCLUSION

If we accept that an enterprise needs to adapt and transform its processes to meet changing output deliverables and customer expectation within an increasingly competitive global environment, we must then have a plan or method by which we could transform the enterprise. Enterprise modelling alone is not sufficient in instigating this transformation. This paper has illustrated the importance of a new progressive enterprise transformation process supported by established enterprise modelling and six sigma methodologies.

The new six sigma driven enterprise transformation process is developed first with a description of an alignment of six sigma with the reference views in generic enterprise architecture and an outline of the framework that encourages the combined benefits of enterprise modelling and six sigma. This generalised framework provides the roadmap for enterprise transformation with readily accepted six sigma tools.

Six sigma focuses on the scope down to incremental improvement projects. Enterprise transformation based on enterprise modelling approach provides a visionary target for the business process engineer to work on. When combining the two methodologies, we can establish the total enterprise picture with the "TO BE" future state, complete with road map and six sigma tools to transform the enterprise can be established.

This proposed approach has been illustrated by a case study. This is an example of how six sigma methodology can be aligned with enterprise modelling to make effective, significant and progressive enterprise transformations. The "TO BE" enterprise model, together with a set of quality criteria as controls, becomes the new operational "AS IS" enterprise model and in the Everest analogy, forms one of the many steps to reaching the summit of enterprise model transformation. Further work is required to detail and standardize the processes of creating effective links between enterprise modelling and six sigma in a unified framework. The six sigma driven enterprise model transformation is a progressive enterprise change process and has proven to induce least disruption to business. It is a less risky approach for enterprise model transformation and improvement.

REFERENCES

1.   Beasley, M.S., Clune, R., Hermanson, D.R. (2005). Enterprise risk management: An empirical analysis of factors associated with the extent of implementation. Journal of Accounting and Public Policy, 24, 521-531

2.   Buhman, C., Kekre, S., Singhal, J. (2005). Interdisciplinary and interorganizational research: establishing the science of enterprise networks. Production and Operations Management, 14(4), 493-514

3.   Chen, D., Doumeingts, G., Vernadat, F. (2008). Architectures for enterprise integration and interoperability- Past, present and future. Computers in Industry, Vol.59, Iss.7, Sept, pp.647-659

4. Corbett L. M. (2008). Manufacturing strategy, the business environment, and operations performance in small low tech firms. International Journal of Production Research. pub. Taylor and Francis. Vol.46, Iss.20, pp.5491-5513
5. Dewhurst, F.W., Barber, K.D., Pritchard, M.C. (2002). In search of a general enterprise model. Management Decision, 40(5), 418-427
6. Goel S., Chen, V. (2008). Integrating the global enterprise using six sigma: Business process reengineering at General Electric Wind Energy. International Journal of Production Economics, Volume 113, Issue 2, June, pp.914-927
7. Goh, T.N. (2002). A strategic assessment of six sigma. Quality and Reliability Engineering International, Vol. 18 No.5, pp.403-410.
8. Jiju, A. (2004). Six sigma in the UK Service Organizations: results from a pilot survey. Managerial Auditing Journal, Vol.19, No.8, pp.1006-1013
9. Jochem, R. (2006). Enterprise Engineering Method supporting six sigma Approach. Proceedings of the 11th World Congress on Total Quality Management, 4-6 December, Wellington, NZ
10. Kutsch E., Hall M., Intervening conditions on the management of project risk: Dealing with uncentainty in information technology projects, International Journal of Project Management, 23, 2005, 591-599
11. Kwon, D., Oh, W., Jeon, S. (2007). Broken ties: The impact of organizational restructuring on the stability of information-processing networks, Journal of Management Information Systems, 24(1):201-231
12. Lin, C.Y.Y., Wei, Y.C. (2006). The role of business ethics in merger and acquisition success: An empirical study, Journal of Business Ethics, 69(1):95-109
13. Love, D., Barton, J. (1996). Evaluation of design decisions through CIM and simulation. Integrated Manufacturing Systems, 7(4), 3-11
14. Marino, A.M., Zabojnik, J. (2006). Merger, ease of entry and entry deterrence in a dynamic model, Journal of Industrial Economics, 54(3):397-423
15. McGinnis, L. F. (2007). Enterprise modeling and Enterprise Transformation. Information knowledge Systems Management 6(1/2), 123-143
16. Mo J.P.T., Zhou M., Anticev J., Nemes L., Jones M., Hall W. (2006). A study on the logistics and performance of a real 'virtual enterprise. International Journal of Business Performance Management, 8(2-3):152-169
17. Molina, A., Medina, V. (2003). Application of enterprise models and simulation tools for the evaluation of the impact of best manufacturing practices implementation. Annual Reviews in Control, 27:221–228
18. Oberg, C., Henneberg, S. C., Mouzas, S. (2007). Changing network pictures: Evidence from mergers and acquisitions. Industrial Marketing Management, 36(7):926-940
19. Ortiz, A., Lario, F., Ros, L. (1999). Enterprise Integration—Business Processes Integrated Management: a proposal for a methodology to develop Enterprise Integration Programs. Computers in Industry, Vol.40, pp.155, 171
20. Rho B. H., Park K., Yu Y. M. (2001). An international comparison of the effect of manufacturing strategyimplementation gap on business performance. International Journal of Production Economics, Vol.70, pp.89-97
21. Rouse, W. B. (2005a). A Theory of Enterprise Transformation. Systems Engineering. Wiley InterScience, Vol.8, Iss.4, pp.279-295

22. Rouse, W.B. (2005b). Enterprises as systems: Essential challenges and enterprise transformation, Systems Engineering, Vol.8, Iss.2, pp.138-150
23. Smith, H., Fingar, P. (2003). Digital six sigma – Integrating continuous improvement, with continuous change, with continuous learning. BPTrends, Pub. Boston University Corporate Education Centre, http://www.bptrends.com, December
24. Smith, L.R., (2001). Six sigma and the Evolution of Quality in Product Development. Six Sigma Forum Magazine, Vol.1, No.1, pp.28-35
25. Török, J. (2004). 1-2-3 Model for Successful Six Sigma Project Selection. Pub. iSixSigma.com. First pub. 20 April 2005. Retrieved December 27, 2008, from http://europe.isixsigma.com/library/content/c050420b.asp
26. Uppington, G., Bernus, P. (2003). Analysing the Present Situation and Refining Strategy. In "Handbook on Enterprise Architecture", Ed. Bernus, P., Nemes, L., Schmidt, G. pub. Springer-Verlag, ISBN 3-540-00343-6, pp.309-332
27. Williams T.J., Bernus P., Brosvic J., Chen D., Doumeingts G., Nemes L., Nevins J.L., Vallespir B., Vlietstra J., Zoetekouw D. (1994). Architectures for integration manufacturing activities and enterprises. Computers in Industry, 24:111-139
28. Yin, X.L., Shanley, M. (2008). Industry determinants of the "merger versus alliance" decision. Academy of Management Review, 33(2):473-491

PART III

IMPLEMENTATIONS

CHAPTER 11

EVALUATING AND REFINING THE "ENTERPRISE ARCHITECTURE AS STRATEGY" APPROACH AND ARTIFACTS

M. DE VRIES and A. C. J. VAN RENSBURG

11.1 INTRODUCTION

Contrary to the information technology and cost reduction foci of previous EA endeavours, this research is used to emphasise a new value-creation focus that includes business architecture and enables business strategy. In support of this new focus, Ross et al. [1] defined a new EA approach that incorporates EA decision-making as part of the strategic decision-making processes of an organisation. Action research is used to gain qualitative feedback on the perceived practicality of two key artefacts that are used to underpin this new approach.

11.2 ENTERPRISE ARCHITECTURE DEFINED

The first traces of EA were found in the publication of Zachman [2]. Zachman [3] defined EA as follows: "Descriptive representations (i.e. models) that are relevant for describing an enterprise such that it can be produced to management's requirements (quality) and maintained over the period of its useful life (change)." Zachman introduced the Zachman framework, which consists of various models that are used to define and communicate six characteristics/abstractions (What, How, Where, Who, When, and Why) for five different viewpoints/perspectives (Planner, Owner, Designer, Builder, and Sub-contractor) (Zachman [3]). The Zachman framework "is a tool for managing and communicating the vast amount of information needed to make broad decisions, those that enable the organisation to be competitive" (O'Rourke, Fishman & Selkow [4]).

Numerous EA definitions were formulated following the inception of the Zachman framework. These definitions addressed the following elements with different emphases:

- Providing a systems view—i.e. describing systems, their components (e.g. people, processes, information, and technology), their interaction, and interrelationships. This includes the use of decomposition strategies to ensure holistic solutions in terms of solution components (TOGAF [5]; Theuerkorn [6]; Gartner in Lapkin [7]; Handler [8]).
- Providing a blueprint for directing the company in terms of required high-level processes and IT capabilities (Ross et al. [1]; Gartner in Lapkin [7]; Boar [9]).
- Defining a process/master plan to explore and model the current realities and the envisioned future state, and enable its evolution (Gartner in Lapkin [7]; Bernard [10]; Schekkerman [11]).
- Defining principles that govern the design and evolution of systems (TOGAF [5]; Theuerkorn [6]; Gartner in Lapkin [7]; Wagter, van den Berg, Luijpers & van Steenberg [12]).
- Using tools, processes and governance structures to implement enterprise-consistent IT architectures (Kaisler, Armour & Vallivullah [13]; Gartner in Lapkin [7]; Schekkerman [11]).

11.3 ENTERPRISE ARCHITECTURE: CREATING GOVERNANCE ON A STRATEGIC LEVEL

EA initially aimed at modelling/describing the architecture components associated with information technology. EA value was limited to direct improvements in the performance of IT itself (lowering overall costs from IT). This approach demonstrated some form of return on investment

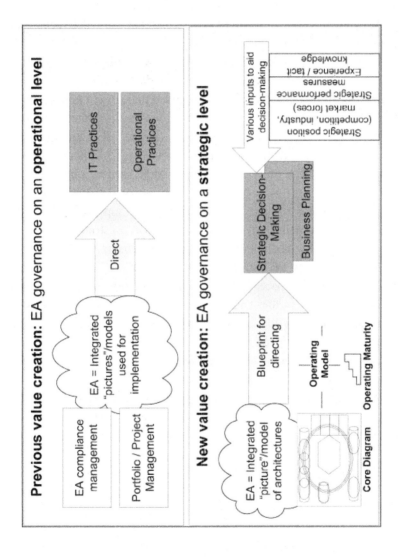

FIGURE 1: Different EA value creation approaches

(ROI)—i.e. accelerating project start-up and decreasing investment in staff, consulting, training, and tools.

Today, EA has broadened from enterprise-wide IT architecture (EWI-TA) to include business architecture (BA); that is, EA = BA + EWITA (Malan & Bredemeyer [14]; Bernard [10]; Ross et al. [1]). The focus is on optimisation "across boundaries to achieve system goals" and the "translation of strategy into implementation" (Malan & Bredemeyer [14]). The change in focus is closely related to the restricted contribution of previous EA value propositions. EA practitioners realised that EA could show more significant value when used to improve business performance, and with IT used to support the execution of strategy (Rosser [15]; Lapkin [16]). Ross et al. [1] took one step further. They believe that EA is not only about supporting the execution of strategy, but should be used as a blueprint for directing strategy.

Figure 1 compares the previous value creation approaches with the new value creation approach of Ross et al.[1]—'EA as Strategy'.

EA should provide a directional blueprint to ensure that companies build a foundation for execution—i.e. they use their IT infrastructure and digitised business processes to automate the company's core capabilities. The rationale is that routine processes are digitised to provide reliability and predictability in business-critical processes. Once these processes have been digitised, management can shift their attention from fighting fires on lower-value activities to strategic issues. Ross et al. [1] recommend eight steps in creating a 'foundation for execution'. During the first three steps, three key artefacts are defined, which should be used in combination to establish EA objectives. The key artefacts are now discussed briefly.

11.3.1 OPERATING MODELS

Ross et al. [1] suggest that organisations should decide on an operating model for the entire organisation on which to build a foundation for execution. The selected operating model provides a "stable and actionable

	Coordination	**Unification**
High	• Shared customers, products, or suppliers • Impact on other business unit transactions • Operationally unique business units or functions • Autonomous business management • Business unit control over business process design • Shared customer/supplier/product data • Consensus processes for designing IT infrastructure services; IT application decisions made in business unit	• Customers and suppliers may be local or global • Globally integrated business processes often with support of enterprise systems • Business units with similar or overlapping operations • Centralised management often applying functional/process/business unit matrices • High-level process owners design standardised processes • Centrally mandated databases • IT decisions made centrally
Low	**Diversification** • Few, if any, shared customers or suppliers • Independent transactions • Operationally unique business units • Autonomous business management • Business unit control over business process design • Few data standards across business units • Most IT decisions made within business units	**Replication** • Few, if any, shared customers • Independent transactions aggregated at a high level • Operationally similar business units • Autonomous business unit leaders with limited discretion over processes • Centralised (or federal) control over business process design • Standardised data definitions but data locally owned with some aggregation at corporate • Centrally mandated IT services

Business process integration (vertical axis, High to Low)

Low   High

Business process standardisation

FIGURE 2: Characteristics of four operating models (Ross et al. [1:29])

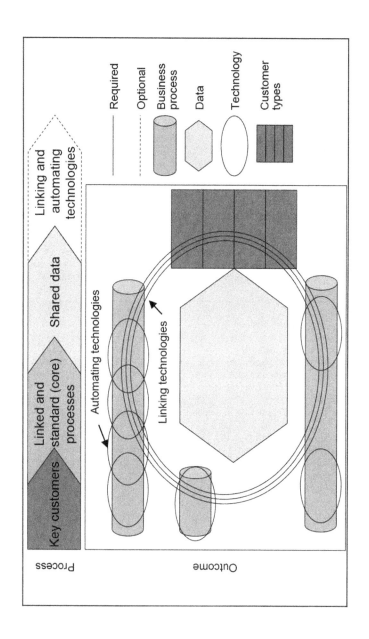

FIGURE 3: Core diagram template for a unification operating model (Ross et al. [1:54])

view of the company" (Ross et al. [1]) and is used to shape future strategic choices.

An operating model is "the necessary level of business process integration and standardisation for delivering goods and services to customers" (Ross et al. [1]). The operating model is a "choice about what strategies are going to be supported", "a commitment to a way of doing business" (Ross et al. [1]). Although each operating model encapsulates numerous characteristics, two key dimensions are used to define four operating models (see Figure 2).

- Business process standardisation, defining how the processes will be executed regardless of the responsible entity or place of execution.
- Business process integration, connecting the efforts of organisational units through linked processes and shared data.

11.3.2 CORE DIAGRAMS

While the operating model defines the process standardisation/integration requirements of the company, the core diagram is used to translate these requirements into the necessary organising logic for business processes and IT infrastructure. The core diagram should be used to:

- Facilitate discussions between business and IT managers to clarify requirements for the company's foundation for execution, and
- Communicate the vision (high-level business process and IT requirements of a company's operating model).

A core diagram contains four main components: (1) core business processes—the stable set of enterprise processes required to execute its operating model and respond to market opportunities; (2) shared data driving the core processes—e.g. customer data shared across product lines or business units of a company; (3) key linking and automation technologies – technologies that enable integration of applications (middleware) to shared data, major software packages such as ERP systems, portals providing standardised access to systems and data, and electronic interfaces to key stakeholder groups; and (4) key customers— major customer groups served by the foundation for execution (Ross et al. [1]). The template for a unification operating model is given in Figure 3.

11.3.3 OPERATING MATURITY ASSESSMENT

Ross et al. [1] also defined four operating maturity levels. In terms of IT management, each level requires different business objectives, IT capabilities, IT funding priorities, and management capabilities. Usually immature organisations will start with a focus on IT efficiency/cost reduction. For instance, research performed by Gartner (Kreizman, Knox & James [17]) indicated that research respondents still ranked IT cost reduction as the most important driver for justifying EA investments. This merely reflected their current operating maturity (stage 1 or 2 according to Figure 4 and its associated cost reduction business objectives. The IT cost-focus of the respondents also correlated with the comparatively low number of business architects (compared with other full-time equivalent architects) employed by the organisation (Kreizman et al. [17]). As organisations mature, business operational efficiency and strategic agility become more important (see Figure 4, stages 3 and 4).

The operating model should be used in combination with the current operating maturity of the organisation to identify realistic EA objectives. If an organisation is, for example, at the first stage of operating maturity, the organisation needs to standardise all technology infrastructure (elevating to stage two) irrespective of the operating model of the organisation. However, elevating from stage two (standardised technology) to stage three (optimised core) requires that the organisation define process standardisation and integration objectives according to the required operating model.

11.4 ENTERPRISE ARCHITECTURE AND STRATEGIC MANAGEMENT

Ross et al. [1] believe that the key artefacts that were discussed in the previous sections could be used to direct the strategic decision-making processes and to shape future strategic choices. Strategic decision-making

	Stage 1: Business Silos	Stage 2: Standardised Technology	Stage 3: Optimised Core	Stage 4: Business Modularity
Business Objectives	ROI of local business initiatives	Reduced IT costs	Cost and quality of business operations	Speed to market; strategic agility
IT Capabilities	Local IT applications	Shared technical platforms	Companywide standardised processes or data	Plug-and-play business process modules
Funding Priorities	Individual applications	Shared infrastructure services	Enterprise Applications	Reusable business process components
Management Capabilities	Technology-enabled change management	Design and update of standards; funding shared services	Core enterprise process definition and measurement	Management of reusable business processes

FIGURE 4: Learning requirements of the architecture stages (adapted from Ross et al. [1:83])

is, according to Johnson, Scholes & Whittington [18]), one of three components of strategic management. The three components consist of (1) defining the strategic position (e.g. current strategic capability, the environment, expectations and purposes); (2) defining strategic choices (e.g. corporate-level/international decisions, business level decisions); and (3) defining strategy execution (e.g. organising, enabling, and managing strategic change). One would require inputs from two of these components (strategic position and previous strategic choices) to define the three key artefacts. The modelled artefacts should then be used in combination to influence the direction of future strategic choices and the subsequent strategic objectives. The set of strategic objectives may then be converted to strategic initiatives/projects with various strategic conversion mechanisms, such as those defined by Kaplan & Norton [19, 20, and 21] (balanced scorecards, strategic themes, and strategy maps). The conceptual process is delineated in Figure 5.

11.5 RESEARCH PURPOSE AND DESIGN

11.5.1 PURPOSE

The purpose of this research was to receive feedback on the perceived practicality of defining the first two key artefacts—the operating model and the core diagram. Action research was used to receive qualitative feedback on the difficulties experienced in defining the current operating model and the core diagram for an organisation/subdivision.

11.5.2 RESEARCH DESIGN

Action research was selected for qualitative research for the following reasons:

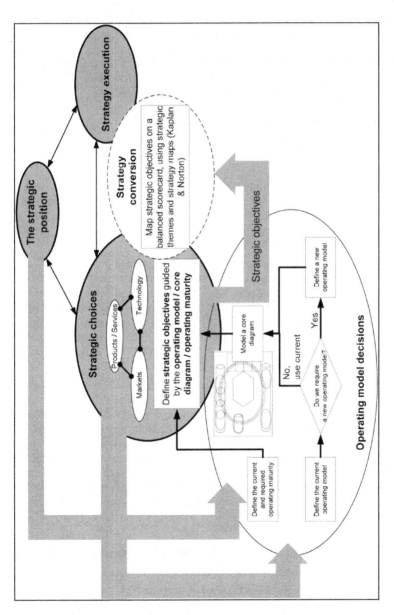

FIGURE 5: Key artefacts contextualised in terms of strategic management components

- The 'EA as strategy' approach of Ross et al. is still new (published in 2006). Research respondents needed to have a good understanding of EA in general and of the new 'EA as strategy' approach. The 'Business Architecture' post-graduate course was used as a vehicle to convey knowledge about EA and the 'EA as strategy' approach, techniques, and artefacts to students, who were then used as respondents.
- The action research process provided the opportunity to assess the students' understanding of the course content, and guide them towards the correct use of the 'EA as strategy' approach, techniques, and artefacts.

The action research process that was followed is based on the work of specialists (referred to by Hodgkinson and Maree [22]):

- Planning—A literature study was conducted in the field of EA to design the course content and assessment mechanisms. Special emphasis was placed on strategic management, the 'EA as strategy' approach, techniques and artefacts, the business architecture domain, and the development of an EA plan.
- Implementation—Live presentations from the course presenter and industry speakers, course notes, and literature references were used to convey the course content to students. Students then had the opportunity to work individually or in pairs and to select an organisation in which to implement some of the techniques presented in the course. An interim project report was submitted for assessment. Students also wrote a semester test to assess their understanding of EA principles and of the 'EA as strategy' approach, techniques, and artefacts defined by Ross et al. [1].
- Observation—The course presenter observed/assessed the students' understanding of the course content. Feedback was given to the students in the light of their semester test and interim project report. Students now had the opportunity to improve/update their project reports and submit a final project report. Based on the final report, they had to submit a completed survey.
- Evaluation – The final reports were assessed and surveys were analysed. Analysis of qualitative survey feedback gave new insight into the practicality of two key artefacts (operating models and core diagrams). New insights were used to define suggested improvements, recommendations, and an agenda for further research.

The survey consisted of twenty-eight questions. Some of the questions were taken from the on-line survey used by the Institute for Enterprise Architecture Developments (IFEAD) (Schekkerman [23]). Categorisation of business activities was taken from the Oracle Magazine subscription form (Oracle Magazine [24]). Questions were categorised according to param-

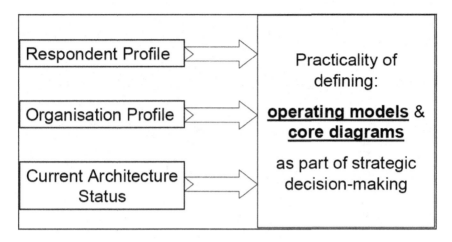

FIGURE 6: Parameters that influence the practicality of defining two key artefacts

eters that could have a significant influence on the perceived practicality of defining the two key artefacts – the operating model and the core diagram (see Figure 6).

11.6 RESULTS

Thirty post-graduate students took part in the final assessment mechanism. As students had the opportunity to work in pairs, a total of twenty-one final project reports and completed surveys was submitted.

11.6.1 STUDENT PROFILE

Figure 7 indicates that fifty-two percent (52%) of the students had previously obtained an engineering degree, thirty-two percent (32%) a technical diploma, twelve percent (12%) a Bachelor of Science (BSc) degree, and four percent (4%) a Bachelor of Commerce (BCom) degree. Tertiary qualifications also correlated with the working positions of the students. Most of the students held positions that were related to business process

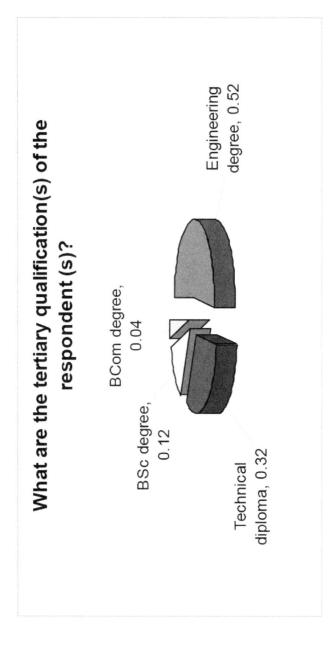

FIGURE 7: Tertiary qualifications of the students

planning and improvement. Questions regarding prior knowledge about information systems indicated that sixty-seven percent (67%) of the students had previously enrolled for information system-related courses, while thirty-eight percent (38%) indicated work-exposure in the field of information systems.

Finally, students had to indicate the main reason(s) for course enrolment. Students could provide additional reasons if the standard categories were not sufficient. Most of the students (19 out of 21) selected the course as part of their Honours studies. They also showed significant interest in the management/improvement of business processes (18 out of 21) and organisational management/governance (12 out of 21).

11.6.2 ORGANISATION PROFILE

Most of the companies that were used for analysis purposes by the students employed fewer than 10,000 employees (see Table 1).

The 21 analysed companies were involved in a wide spread of 19 different business activities. Note that a company could be involved in multiple business activities. These included automotive manufacturing (5 out of 21), the consumer sector (4 out of 21), high- technology manufacturing OEM (3 out of 21), industrial manufacturing (3 out of 21), professional services (3 out of 21), research (3 out of 21),other business services (5 out of 21) and 12 remaining business activities (17 companies out of 21). None of the analysed companies was in the financial/insurance services industry. According to Matthee, Tobin & Van der Merwe [25], the financial sectors usually invest in EA endeavours owing to their high dependency on IT.

11.6.3 CURRENT ARCHITECTURE STATUS

Figure 8 indicates that a large number of companies (9 out of 21) managed their divisions in silos. A significant number had progressed to the level of standardised technology (7 out of 21) and optimised core (5 out of 21). None of the companies operated according to a modular business design.

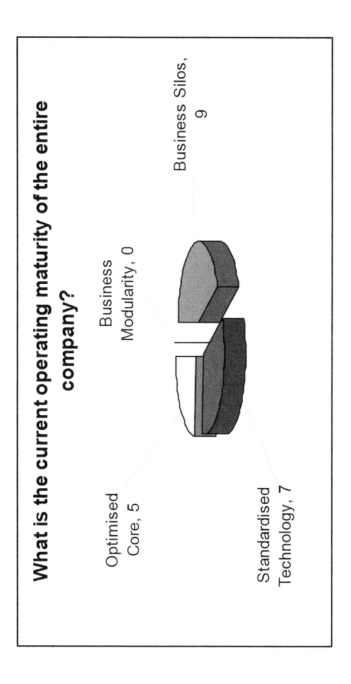

FIGURE 8: Operating maturity of the companies

According to Table 2, business architecture was well-established at 11 out of 21 companies. The perceived level of business architecture activity may also be explained by the process inclination of the students.

TABLE 1: Size of the entire company

Number of people working in the organisation	Number of companies
96,000	1
33,000	1
10,000–24,999	2
100–9,999	12
1–99	5

EA governance activities were performed at thirty-eight percent (38%) of the analysed companies. Students believed that a company should invest in EA governance owing to its decision-making support (7 out of 21), system development support (6 out of 21), and delivery of insight and overview of business & IT (5 out of 21). Only four students indicated the use of architecture modelling technology that includes a repository. Tools include ARIS, Casewise, and Systems Architect. According to Figure 9, eight companies (38%) did not use a framework.

11.6.4 THE PERCEIVED PRACTICALITY OF OPERATING MODELS AND CORE DIAGRAMS

11.6.4.1 THE ANALYSIS LEVEL FOR DEFINING AN OPERATING MODEL

Students preferred to apply the 'EA as strategy' approach on a business unit level (17 out of 21) rather than a corporate level (4 out of 21). The different types of operating models were well-represented: diversification (7 out of 21), unification (6 out of 21), replication (5 out of 21), and coordination (3 out of 21).

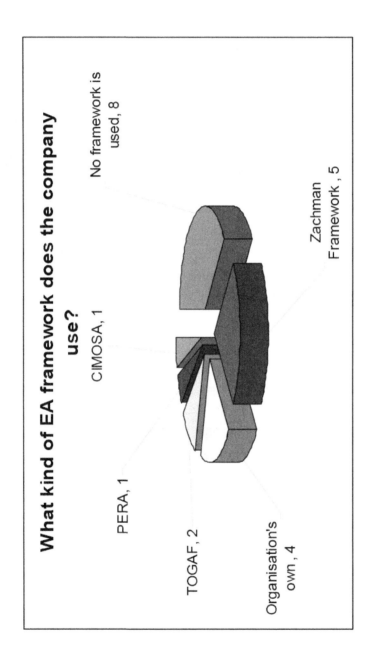

FIGURE 9: Enterprise architecture frameworks in use

11.6.4.2 DIFFICULTIES IN DEFINING THE CURRENT OPERATING MODEL

Students indicated their difficulty in deciding on one specific operating model (14 out of 19). A few students (4 out of 19) indicated minimal difficulty in identifying the operating model. Qualitative feedback was tagged according to emerging themes:

- Difficulty was experienced in deciding on a single operating model (8 out of 14 who experienced difficulty). Students had difficulty in establishing the degree of process standardisation/integration that would be required to classify an organisation according to a specific model. Companies (especially on a corporate analysis level) exhibited behaviours of multiple operating models.
- Students (5 out of 14 who experienced difficulty) conveyed their difficulty in finding the correct information to perform a classification. This was also attributed to the limited knowledge and awareness of EA in the company.
- Some difficulty (1 out of 14 who experienced difficulty) occurred in defining an operating model on a business unit level due to fuzzy boundaries between the corporate level and business unit level.

TABLE 2: Established architectural levels

Architecture Levels	Number of companies
Business Architecture	11
Information-System Architecture (Applications Architecture)	7
Enterprise Architecture	6
Security Architecture	6
Information Architecture	5
Technology Infrastructure Architecture	5
Governance Architecture	3
Software Architecture	3

11.6.4.3 DIFFICULTIES IN COMPILING A CORE DIAGRAM

Qualitative feedback was tagged according to emerging themes:

- Half the students (10 out of 20) experienced difficulty in selecting the main components of the core diagram. These students had trouble in identifying the shared technologies (4 out of 10 who experienced difficulty), shared data (3 out of 10), shared processes (3 out of 10), and the key customers (1 out of 10). The problematic identification of shared technologies may be attributed to the student profile/limited exposure to technology infrastructure.
- Some students (6 out of 20) had difficulty in understanding the generic core diagram templates provided by Ross et al. [1] or relating the diagram components to their company. They also questioned the validity of their own core diagram designs.
- Another concern was the availability and/or the consolidation of available information (4 out of 20 students).

11.7 SUMMARY AND INTERPRETATION OF RESULTS

It was found that most of the students had an engineering background, held positions related to business process planning and improvement, and showed significant interest in the management/improvement of business processes and organisational management/governance. Students also had sufficient knowledge of information systems.

Concerning the organisation profile, most of the companies that were used for analysis purposes employed fewer than 10 000 employees, and were involved in a large number of business activities excluding the financial sector. Results further indicated a relatively low level of operating maturity – most of the analysed companies displayed business silo behaviour, while none of the companies operated according to a modular business design. The study indicated that business architecture was well established at the analysed companies. Use of architecture modelling technology was limited.

The perceived practicality of the operating model and core diagram artefacts could not be evaluated on a corporate level, as most of the students defined operating models at a business unit level. According to Ross et al. [1], this should not be a hurdle in validating the artefacts per se, as operating models and core diagrams may be defined at various levels of the organisation. The interpretation of the various difficulties experienced follows:

- Difficulty in selecting a single operating model is linked to the identification of the degree of process standardisation/integration for the analysed organisation/business unit. Extensive implicit/explicit knowledge is implied during the evaluation of the operating model characteristics that define the degree of process standardisation/integration.
- Students had difficulty in finding the correct information to perform an operating model classification or select core diagram components. Identification of operating model characteristics and core diagram components requires knowledge about the strategic choices (markets, products/services), operating/organising logic, business processes, and main databases and technologies of the organisation. Some baseline architectures are thus required, and this knowledge is not necessarily available or in an explicit format.
- Students experienced difficulty in selecting the main components of the core diagram and understanding the core diagram templates. This may be related to the limited set of examples provided in the textbook. Case studies would be required to demonstrate inputs that would be required (e.g. baseline architectures) to define the core diagram components.

11.8 CRITICAL EVALUATIONS AND INFERENCES

Based on the qualitative feedback received from the action research effort, the researcher revisited the main objectives of the operating model and core diagram:

- To aid the main stakeholders/users of these artefacts (business and IT managers) in guiding them during their strategic decision-making processes.
- To communicate architecture vision to other stakeholders (in terms of process standardisation/integration requirements).

If the main stakeholders are to use these artefacts to guide them during the strategic decision-making processes, the artefacts should be based on a more rigorous approach to attaining the artefact outputs. This will increase their validity and reliability. The researcher also believes that process standardisation/integration requirements should be based on a more scientific approach to define optimal standardisation/integration requirements for an organisation. Porter [26], for instance, believes that decisions regarding process standardisation/integration are complex and require detailed analysis based on the strategic intent of the organisation (e.g. cost lead-ership/differentiation/ focus-driven for target segments). Cost leadership companies, for instance, would have to assess the impact that process stan-dardisation/integration could have on overall cost, while differentiation-focused companies need to assess if process standardisation/integration could increase the uniqueness of an activity or lower its cost of differen-tiation.

11.9 CONCLUSIONS

This study emphasised the limited value gained from EA when measured in terms of ROI due to cost reductions alone. Today EA practitioners realise that new value propositions emerge when EA is used to support the strate-gic direction of the organisation. This new focus was used to introduce a new approach towards EA value creation, called 'EA as strategy'. The ap-proach incorporates EA planning as part of the strategic decision-making process using three key artefacts: operating models, core diagrams, and an operating maturity assessment.

Action research was used to assess the practicality of two key artefacts (operating model and core diagram), which highlighted some difficulties that were experienced and led to some critical evaluations and recommen-dations regarding the artefacts. It is believed that the operating model and core diagram could be useful in visualising the process standardisation/ integration requirements of an organisation/sub-division. The artefacts should, however, be supported by a more scientific approach to their deri-vation, to increase their validity/reliability.

Further research has been initiated to perform a case study at an or-ganisation. The case study incorporates processes to model baseline archi-tectures, current strategic choices (markets, products/services), operating/ organising logic, business processes, main databases, and technologies of the organisation. This will be followed by various analyses (e.g. value chain analyses) to identify process standardisation/integration opportuni-ties. Current artefact designs (e.g. operating model and core diagram) may need to be adapted to convey the process standardisation/integration re-quirements to strategic decisionmakers. The new artefact designs will be distributed to different strategic decision-makers to gain feedback about their usefulness during strategic decision-making.

REFERENCES

1. Ross, J.W., Weill, P. and Robertson, D.C. 2006. Enterprise Architecture as strategy. Creating a foundation for business execution. Boston, Massachusetts: Harvard Business School Press.
2. Zachman, J.A. 1987. 'A framework for information systems architecture', IBM Systems Journal, 26(3), pp. 276-292.
3. Zachman, J.A. 1996. 'Enterprise Architecture: The issue of the century'. [online] URL: http://mega.ist.utl.pt/~ic-atsi/TheIssueOfTheCentury.pdf. Accessed on 13 November 2007.
4. O'Rourke, C., Fishman, N. and Selkow, W. 2003. Enterprise Architecture using the Zachman Framework. Boston, Massachusetts: Thomson Course Technology.
5. TOGAF Version 8.1, enterprise edition. [online] URL: www.opengroup.org/architecture/togaf8/downloads.htm#Non-Member. Accessed 13 November 2007.
6. Theuerkorn, F. 2005. Lightweight enterprise architectures. New York: Auerbach Publications.
7. Lapkin, A. 2006. 'Gartner defines the term "enterprise architecture"', Gartner Research, ID Number: G00141795.
8. Handler, R. 2004. Enterprise Architecture is dead – long live Enterprise Architecture. http://itmanagement.earthweb.com/columns/article.php/11079_3347711_1. Accessed 5 Jan 2007.
9. Boar, B. H. 1999. Constructing blueprints for enterprise IT architecture. New York: J. Wiley.
10. Bernard, S.A. 2005. An introduction to Enterprise Architecture EA3. 2nd edition. Bloomington: AuthorHouse.
11. Schekkerman, J. 2004. How to survive in the jungle of Enterprise Architecture frameworks. 2nd edition. Victoria: Trafford Publishing.
12. Wagter, R., Van den Berg, M., Luijpers, J. and Van Steenberg, M. 2005. Dynamic Enterprise Architecture – How to make it work. Hoboken, New Jersey: John Wiley & Sons, Inc.
13. Kaisler, S.H., Armour, F. and Valivullah, M. 2005. Enterprise architecting: Critical problems, Proceedings of the 38th Hawaii International Conference on System Sciences, IEEE CS Press: 224b.
14. Malan, R. and Bredemeyer, D. 2005. 'Enterprise Architecture as strategic differentiator', Cutter Consortium Executive Report, 8(6), pp. 1-23.
15. Rosser, B. 2004. 'Sell the value of architecture to the business', Gartner Research, ID Number: G00123412.
16. Lapkin, A. 2005. 'Business strategy defines Enterprise Architecture value', Gartner Research, ID Number: G00129604.

17. Kreizman, G., Knox, M. and James, G.A. 2005. 'IT cost reduction and performance improvement provide the most value for Enterprise Architecture', Gartner Research, ID Number: G00130524.

18. Johnson, G., Scholes, K. and Whittington, R. 2005. Exploring corporate strategy. London: Prentice Hall.

19. Kaplan, R.S. and Norton, D.P. 1996. The balanced scorecard. Boston, Massachusetts: Harvard Business School Press.

20. Kaplan, R.S. and Norton, D.P. 2004. Strategy maps: converting intangible assets into tangible outcomes. Boston, Massachusetts: Harvard Business School Press.

21. Kaplan, R.S. and Norton, D.P. 2006. Alignment – using the balanced scorecard to create corporate strategies. Boston, Massachusetts: Harvard Business School Press.

22. Hodgkinson, C.A. and Maree, J.G. 1998. 'Action research: some guidelines for first-time researchers in education', Journal of Education and Training, 19(2), pp. 51-65.

23. Schekkerman, J. 2006. Enterprise Architecture survey 2006, Institute for Enterprise Architecture Developments (IFEAD).http://enterprise-architecture.info/EA_Survey_2006.htm. Accessed 6 November 2007.

24. Haunert, T. 2008. 'Want your own FREE subscription?', Oracle Magazine, June/July, p. 64.

25. Matthee, M.C., Tobin, P.K.J. and Van der Merwe, P. 2007. 'The status quo of Enterprise Architecture implementation in South African financial services companies', South African Journal of Business Management, 38(1), pp. 11-23.

26. Porter, M.E. 2004. Competitive advantage – creating and sustaining superior performance. New York: Free Press.

CHAPTER 12

AN ENTERPRISE ARCHITECTURE METHODOLOGY FOR BUSINESS-IT ALIGNMENT: ADOPTER AND DEVELOPER PERSPECTIVES

ZULKHAIRI M.D. DAHALIN, RAFIDAH ABD RAZAK, HUDA IBRAHIM, NOR IADAH YUSOP, and M. KHAIRUDIN KASIRA

12.1 INTRODUCTION

Enterprise Architecture (EA) can be viewed as a strategic approach in the evolution of the IT system in response to the constantly changing needs of the business environment (Schekkerman, 2006). There is no consensus on the definitions and description of EA. A common theme in all of the definitions is that EA describes principles and guidelines in governing the implementation of information, technology and business mission in organizations; involving different stakeholders and processes.

Enterprise Architecture is a blueprint for how an organization achieves the current and future business objectives using IT. It examines the key business, information, application, and technology strategies and their impact on business functions (Pereira and Sousa, 2005). It provides the framework for planning and implementing a rich, standards-based, digital information infrastructure with well-integrated services and activities (Watson, 2000).

Organizations are always looking to find new and cost effective means to leverage existing investments in IT infrastructure and incorporate new capabilities to improve business productivity (Patrick, 2005). Hence, there is an increasing need for organizations to align their IT and business strategies. This paper examines the Systemic Enterprise Architecture Methodology (SEAM) developed by Wegmann (2003) to determine its relevance in explaining the business-IT alignment. Business-IT alignment can be defined as the adoption of appropriate IT solutions that meets the business requirements and gives satisfactory returns on the IT investment.

12.2 OBJECTIVES

This paper is set up to meet the following objectives: (1) to examine the trend and status of EA adoption and implementation in Malaysia based on international benchmark; and (2) to provide evidence of the significance of the Systemic Enterprise Architecture Methdology (SEAM) as a viable approach in validating business-IT alignment.

12.3 ENTERPRISE ARCHITECTURE IN MALAYSIA

In Malaysia, perhaps the first known publishable article on EA appeared in a book written by Simon Seow (Seow, 2000). Ever since then and through series of workshops and seminars, as well as the setting up of the Malaysia's Chapter for the International Association of Software Architecture (IASA) in 2002, EA is becoming more and more popular among organizations based on the keen interest on the subject and the overwhelming participation among key IS players (Zulkhairi et al., 2006). However, there is still a strong need for academic involvement particularly in research and development of EA in Malaysia to further enrich the knowledge of EA.

A study conducted in 2006 on the practices of EA in selected organizations in Malaysia reveals that knowledge and understanding of EA among the organizations are poor though there had been efforts at implementing EA (Zulkhairi et al., 2006). A study by Rafidah et al. (2007) found that organizations in Malaysia, both public and private, do practice EA but the EA activities were found to be incomplete or not adequately addressed. The authors also found that knowledge on EA is very poor among the enterprise management in Malaysia. In terms of EA practice, the findings suggest variation of EA particularly at the planning stage. The study also reveals that some aspects of the EA framework were not addressed at all; whilst other aspects that were addressed vary in terms of perspectives. Earlier, Seow (2000) observed that actual EA practice among Malaysian organizations was very minimal.

12.4 THE STUDY

Theoretical framework is a deductive reasoning approach where existing theories, ideas, constructs and methodologies are combined in search for relevant explanation to the phenomenon being studied. SEAM is based on business/IT alignment market, in which supplier business systems compete to provide a value to an adopter business system. Two units of analysis were identified in this study. First, those who are responsible for business-IT alignment (the EA Adopter); and second were those who care about EA (the EA Developer). These are people who plan, implement, advice and do consulting and collaborate with others for the development of EA in the organization. The role of the respondents in the EA Adopter is to adopt EA. The supply role in the EA Development can be broken down into two main actions: planning and implementation. The adoption action is mainly the responsibility of managers and staff at the operational level that drives the improvement of the business process. EA Development started with planning, which is the responsibility of senior management, and made practical through implementation, which is essentially the IT professionals. These three actions: Planning, Implementation and Adoption according to Wegmann (2003) are referred to as the EA lifecycle activities. Three groups of respondents were identified in this study to commensurate with these three actions that signify the EA activities.

Elements of the research to be studied are based on the Trends in Enterprise Architecture 2005 report by the Institute for Enterprise Architecture Development (Schekkerman, 2005). EA activities refer to the environment in which EA is present and there is evidence to suggest business-IT alignment exists through interactions of elements between business issues and the EA environmental elements. In this study, these interactions were identified based on correlation analysis that attempts to relate the EA environmental elements with the EA business issues. Relationships that are found to be significant are deemed to have supported the interactions, thereby providing evidence of business-IT alignment. The IFEAD 2005 report presented three components that make up the EA environmental elements. These are the EA Environment, the EA Governance, and the EA Methods, Tools and Framework. These three elements along with the EA business issues were incorporated into the questionnaire design as instrument used to carry out the study. A preliminary study was conducted to test the instrument and was found to be valid (Rafidah et al., 2009).

The two units of analysis mentioned in SEAM, the Adopter and the Developer, were identified as respondents in this study. EA Adopters were those users at the managerial and operational level responsible for the business-IT alignment. EA Implementers represent respondents who plan and implement the EA in the organization. This can be further subcategorized into the Planner, who are essentially the CIO, Chief Architect and IT Manager, and the Implementer, who are the Architect, Consultant and Systems Analyst.

Data collection was based on a questionnaire constructed to fulfill the needs of the two units of analysis, whereas feedbacks obtained followed the construct developed by the IFEAD 2005 report. The IFEAD report, edited by Jaap Schekkerman, President of IFEAD, consists of four dimensional constructs as represented in Table 1. The first construct, the EA Business Issues describes the respondents' perception on the business issues that EA can help addressed. Two questions were posed to operationalize the construct: 1) Why EA is important?; and 2) What business issues can EA help to address? Table 1 lists the complete operationalization of the four constructs adopted from the IFEAD report.

TABLE 1: Dimensional Construct of EA Usage (adopter) and Implementation (Development)

Dimension	Operationalized Research Elements
EA Business Issues	Why is EA important for your organization? For what kind of issues do you plan an EA program?
EA Environment	Is your organization familiar with the importance of EA? Is EA part of your organization's strategic governance? Are you aware of any guidelines or policies related to EA in Malaysia? Is there any architecture established in your organization?
EA Governance	At which level is EA part of your organization's structure? Do you have your own architect? What type of architect do you have? Does your organization use external architect? From which external organizations do you get support in your EA projects? To whom is the architect reporting? How are your architects educated/trained? Is certification of EA by an official authority an issue? How often do you plan your people to coach by experienced architects? How do you select a good architect coach/mentor? How do you get more information about EA?
EA Methods, Tools and Framework	What kind of EA framework does your organization used? What kind of tools you use to develop EA? What kind of business modeling techniques is your organization using? What kind of system modeling techniques is your organization using? What kind of system development methodology is in use in your organization?

EA Environment refers to the situation within the organization that makes EA present possible. EA Governance refers to the structure in which EA is being managed, including the level in which EA is positioned within the organization, the personnel involved, support structure, skills and training involved, and EA knowledge acquisition. The last construct, EA Methods, Tools and Frameworks, is concerned with the organization's adoption of a particular EA framework, the kinds of tools used to develop EA, modeling techniques used, and systems development methodology used to develop information systems that are part of the organization's EA implementation.

This study involves a sample size of 100 organizations from both public and private sectors. The sampling frame was based on the list of organizations registered in the Universiti Utara Malaysia (UUM's) University Industry Link database directory that lists more than 1260 organizations participated in the student practicum attachment throughout Malaysia. In addition, samples were also drawn from the Malaysia Computer Industry

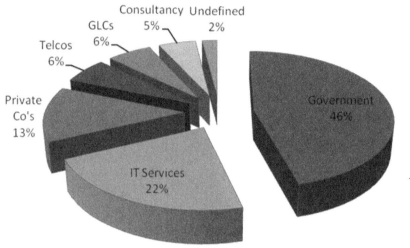

FIGURE 1: Categories of Organizations

Association (PIKOM) directory, Malaysia National Computer Confederation (MNCC), MSC status companies, Federal and State Government, and IASA. Data collection involved three stages which are online, postal, and hand-delivered due to the poor response encountered in the earlier stages. A total of 500 questionnaires were distributed from the list based on random selection with 100 returns representing 20% response rate.

12.5 ORGANIZATIONS BACKGROUND

Figure 1 shows the categories of organizations participated in the survey. Organizations from multinational to small organizations participated in the survey. Majority of the participating organizations (84%) were with less than 1000 people working in the organizations.

The participating organizations were from Kuala Lumpur (26%), Johore (18%), Selangor (11%), Kedah (11%) and smaller percentages (3-6%)

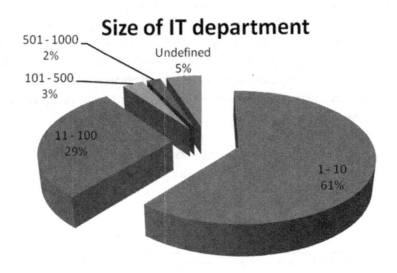

FIGURE 2: Number of Workers in IT Dept.

from other states. Majority are from Government organizations (46%). Other participating organizations were IT Services, Private companies, Telecommunication companies (Telcos), Government-linked (GLCs), and Consultancy firms.

The number of people working in IT department is presented in Figure 2. Sixty-one percent of the participants in the categories of 1-10 people, 29% in the categories of 11-100 people, 3% in the categories of 101-500 and 2% made up the 501-1000.

12.6 THE RESPONDENTS

Table 2 listed the participating organizations category of respondents in the EA lifecycle activities. Majority of the respondents (59%) are in the EA Implementer category, 20% of them are in EA Adopter category and 17% are in EA Planner category. Recall that EA Implementer are those IT professionals and technical people involved in implementing IT solutions that support the business-IT alignment, whilst EA Adopter is essentially the end-users who are managers and operational staff. EA Planner represents the senior level management involved in formulating the business plans and strategies. These categories of respondents were identified based on a cover letter sent to the organizations specifically requesting respondents who were familiar with the organization's IS and business processes to complete the questionnaire. The purpose is to ensure that those who are in the position to represent the organization in terms of EA knowledge and practices should complete the questionnaire.

TABLE 2: Category of Respondents

Category	Freq	%
EA Adopter (End-users and business managers)	20	20
EA Planner (CIO, IT managers, chief architects)	17	17
EA Implementer (IT professionals)	59	59
Undefined	4	4

12.7 ENTERPRISE ARCHITECTURE ACTIVITIES

This section presents evidence of EA activities found in the study. A comparitive analysis is also carried out against an international study that was carried out and reported by the Institute for Enterprise Architecture Development (IFEAD, 2005). Consistent with the report and construction of the questionnaire instrument in this study, EA activities are categorised and presented in the following manner: (1) EA business issues; (2) EA environment; (3) EA governance; and (4) EA methods, tools and frameworks. However, there is a need to include additional factors into the EA categories in view of the dynamic nature of IT and the global business transformations that exist today. Factors such as Business-IT alignment, Customer Satisfaction, Better Work Environment, Improved Project Management, and Service-Oriented Architecture not included in the IFEAD study were found to be important as presented in the sections that follow.

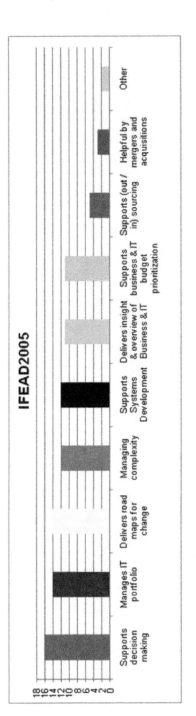
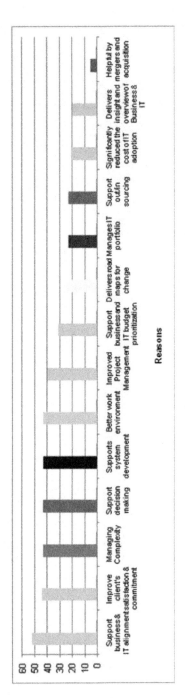

FIGURE 3: EA Important to Organization

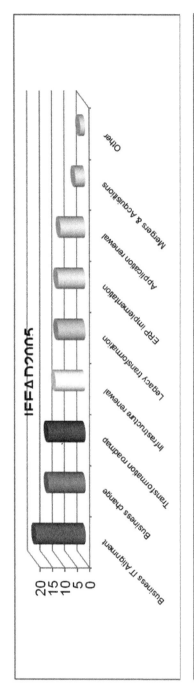

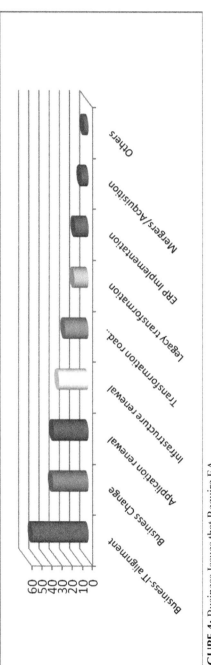

FIGURE 4: Business Issues that Require EA

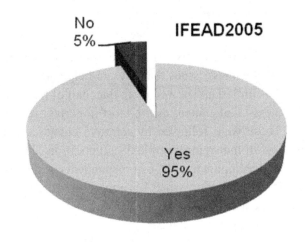

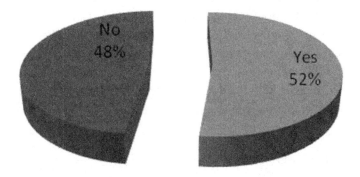

FIGURE 5: Familiar with EA

12.8 EA BUSINESS ISSUES

As mentioned previously, EA Business Issues describes the respondents' perception on the business issues that EA can help addressed. The question on why EA is important showed more than half of respondents (52%) perceived business-IT alignment as the most important reason for EA to organization. These were followed by improve client's satisfaction and commitment (44%), managing complexity, support decision making and support systems development at 43% of respondents, respectively. Figure 3 shows the rest of the reasons why EA is important and a comparison with the IFEAD 2005 report.

The IFEAD 2005 findings indicate support decision making, manages IT portfolio, and delivers road maps for change as the top 3 reasons why EA is important. These were also present in the top 10 list found in this study.

On the kind of business issues that requires EA, again business-IT alignment appeared top with more than half of the respondents (55%) perceived it as most important. This is followed by business change (35%), application renewal (34%), infrastructure renewal (29%), and transformation road map (23%). The rest are found in Figure 4 along with the IFEAD 2005 findings.Comparison with the IFEAD 2005 shows most of the business issues that requires EA are similar across the two studies. Business-IT alignment appears to be a universal issue that requires EA. Similarly business change, application and infrastructure renewal, and transformation roadmap all deals with the dynamic nature of business where respondents believe EA should be able to address.

12.9 EA ENVIRONMENT

EA Environment as described previously refers to the situation within the organization that makes EA present possible.It deals with familarity of the organization with EA, policies and guidelines on EA implementation, and the presence of EA. Finding suggests slightly more than half (52%) of respondents acknowledged that their organizations are familiar with the importance of EA. In contrast, the IFEAD 2005 report shows that almost

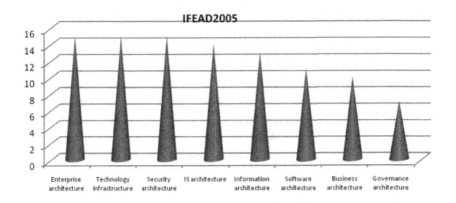

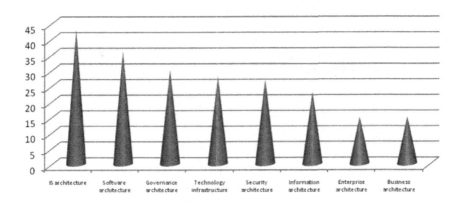

FIGURE 6: Kinds of Architecture Established

all (95%) of the responding organizations are familiar with the importance of EA. Figure 5 presents the comparison of the two studies.

Part of the studying the EA environment is to examine the commitment by organizations to establish some form of architecture. Hence respondents were asked whether they have any architecture adopted by their organizations. From the survey, 43% of the organizations have established Information Systems architecture, which was the most popular kind of architecture indicated by the respondents. Next is Software architecture at 36%, followed by Governance Architecture (30%), and Technology Infrastructure Architecture (28). The rest are presented as in Figure 6 along with a comparison with the findings reported by IFEAD 2005. Surpris-

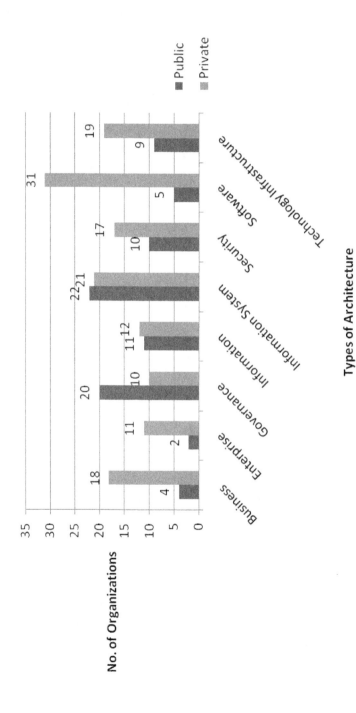

FIGURE 7: Cross-tabulation of Architectures Established in Public and Private Sectors

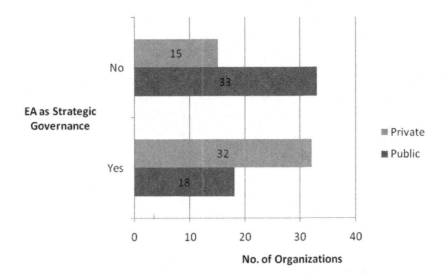

FIGURE 8: EA as Organization's Strategic Governance

ingly, this study found EA at the bottom of the list, as opposed to the IFEAD 2005 report which ranked EA at the top along with Technology Infrastructure and Security architectures.

A cross-tabulation between the public and private sectors shows a remarkable difference in terms of architecture preference. Private sector identified Software architecture as the most dorminant architecture established, however, the public sector indicated Software architecture to be among the least. Equally surprising, EA was at the bottom of the list indicated by the public sector organizations. The public sector identified IS architecture and Governance as the two kinds of architectures mostly adopted. Figure 7 presents the kinds of architectures established between the public and private sectors.

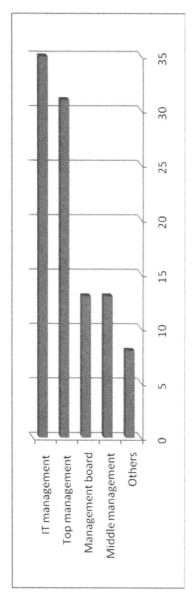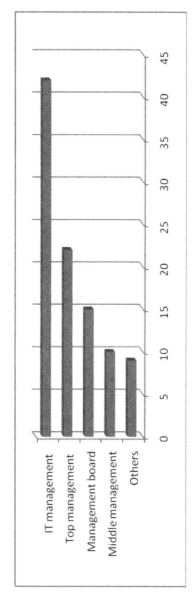

FIGURE 9: Level of EA Governance Structure

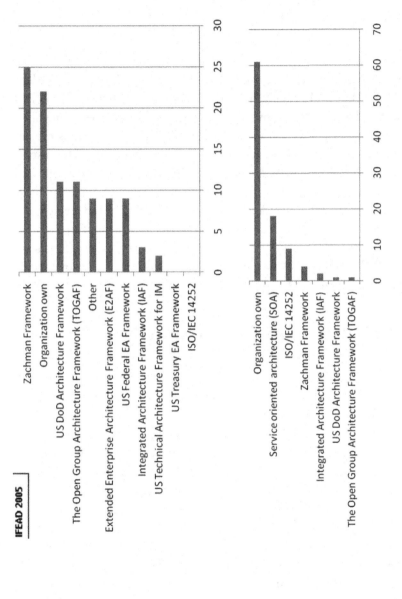

FIGURE 10: Adoption of EA Framework

12.10 EA GOVERNANCE

EA Governance describes the management structure which reflects the conscious efforts place by the organizations in the development and adoption of EA. This study found that 50% of the organizations indicate that EA is part of their strategic governance. A cross tabulation between public and private sectors show that the private sector organizations took greater efforts in making EA part of their strategic governance. Figure 8 presents the findings based on cross-tabulation of EA Governance by public and private sectors.

In terms of the level of EA governance structure, IT Management appears to be the preferred choice with 42% of respondents indicate EA governance structure is at their IT Management level. This is also consistent with the IFEAD 2005 report that shows similar order of EA governance structure as presented in Figure 9.

12.11 EA METHODS, TOOLS AND FRAMEWORKS

As has been described previously, EA Methods, Tools and Frameworks, is concerned with the organization's adoption of a particular methods, tools and framework for the development and adoption of EA. Findings from this study suggest majority of the respondents indicated using their organization's own EA framework with 61% responses. The Service Oriented Architecture (SOA) came a far second with 18% responses. This was followed by the ISO/IEC 14252 standard architecture (IEEE 1003.0) with 9%, and Zachman Framework with 4% responses.

Figure 10 presents the EA Framework adopted by the responding organizations along with the IFEAD 2005 findings. It is interesting to note that the presence of a well-known EA framework like the Zachman Framework is declining dramatically compared to the 2005 report on the trends in EA (IFEAD, 2005). What we have seen at least in Malaysia is that EA framework is still dominantly "home grown" as organizations are still grappling with the idea of an industry standard architecture that can explain how their information systems can support the organization's business objec-

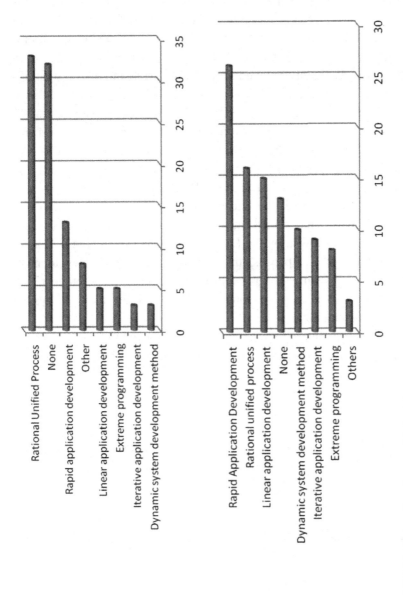

FIGURE 11: System Development Methodology

tives. One industry standard architecture that appears promising and has recently shown significant presence is the Service Oriented Architecture (SOA). In the IFEAD report 2005, SOA was not even mentioned. Today, five years down the road, SOA has become a prominent industry standard architecture and its popularity and adoption is expected to improve significantly in the future as more and more major IT players like IBM, Microsoft, SAP, Oracle etc. incorporate SOA in their service delivery.

Respondents were also asked on the methodology used for systems development. Top of the list is Rapid Application Development (RAD) with 26% of respondents indicated using the RAD approach to develop systems. The Rational Unified Process (RUP) came in second with 16% responses. A close third is the traditional system development life cycle represented by the Linear Application Development (LAD) with 15% responses. This is followed by the Dynamic System Development Method (DSDM) with 10% responses, the Iterative Application Development (IAD) with 9% responses, Extreme Programming (8%), and others (3%). A significant proportion of respondents (13%) did not used any system development methodology.

Figure 11 summarized the System Development Methodology in the participating organizations with comparison from the IFEAD 2005 results. Both RUP and RAD appears to be popular occupying the top 3 development methodology for EA. However, the IFEAD 2005 study is more surprising as a significant proportion of their respondents did not use any known methodology.

As for the tools to develop EA in the organizations, majority reported Microsoft Office Tools (60%) were used to develop EA in their organizations. Almost 30% indicated using Microsoft Visio.

12.12 SEAM VALIDATION

With evidence of EA activities presented in the foregoing section, recall that SEAM is an EA methodology proposed by Wegmann (2003) describing the business-IT alignment market in terms of the supplier business system collaborate with the adopter business system in the form of EA lifecycle activities. This section attempts to present evidence that the SEAM

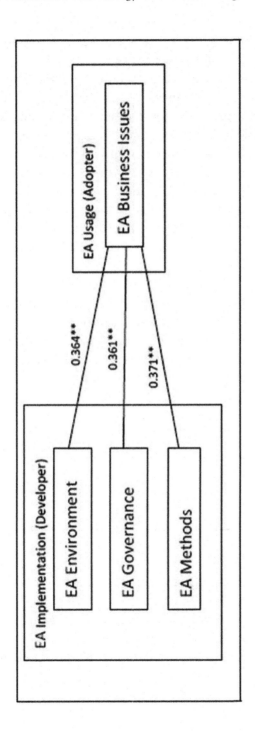

FIGURE 12: SEAM Validation Correlation Analysis

is a valid approach to EA development and adoption. Applying SEAM to this research, the supplier business system is represented by the EA Developer, which as mentioned in Section 4 of this paper, consists of 2 units of analysis: EA Planner and EA Implementer. The adopter business system is represented by the third unit of analysis, ie., EA Adopter. The collaboration between the EA Developer and EA Adopter in the business-IT alignment market can be illustrated by the extend relationships exist between these two elements. Using the dimensional construct of EA usage (adopter) and EA implementation (developer) as presented previously in Table 1 to signify the EA lifecycle activities, correlation analysis can be performed to determine the madnitude of relationships exist between the constructs. These relationships are presented as shown in Figure 12.

In order to test the significance of the relationships between the constructs, Intensity indices were formulated for each construct to determine the strength of the construct based on selections made by the respondents on the questionnaire instrument. Average scores of the intensity indices along with minimum and maximum scores were calculated and presented as in Table 3.

TABLE 3: Intensity Indices of EA Constructs

Intensity Construct	Average (%)	Minimum (%)	Maximum (%)
Business Issues	28.8	4.17	83.3
EA Environment	33.0	0.00	100.0
EA Governance	22.0	0.00	61.9
EA Methods, Tools & Framework	17.0	5.26	36.84

Table 3 suggests that the strengths of EA activities are determined largely by the EA environment, that is familiarity of the organization to EA, the presence of EA policies and guidelines, and actual implementation of EA. EA activities are also determined by intensity of business issues, that is the need to address business changes and transformation processes. To a lesser extend, EA Governance also affect EA activities but not as much as EA Environment and Business Issues. The least average intensity score for EA Methods, Tools and Framework suggests that implementation issues, particularly the technical development of EA is less prominent than the business issues. This is despite the sample of this study constitutes nearly 60% of respondents were categorised as EA Implementers.

Figure 12 shows the results of the correlation analysis between the EA Implementation (developer category constituting the EA Planner and EA Implementer) and the EA Usage (adopter category). The results suggest there exist relationships between EA Environment, EA Governance and EA Methods, Tools and Frameworks with EA Business Issues at the 0.01 significant level. Correlation coefficients reveal the magnitude and direction of relationships. The magnitude or degree of correlation between 0.36 to 0.37 as shown in the figure is considered modest. The positive relationships between the variables indicate a large (or small) values on the Developer category are associated with a large (or small) values on the Adopter category. In other words, the higher the intensity of EA Environment as signified by familiarity with EA, EA as part of organization's strategy, number and variety of architectures, and EA policy and guidelines, the higher the intensity of business issues addressed by EA. Similarly, the higher the intensity of EA Governance signified by EA formal structure, presence and levels of architects, and architects' reporting structures, the higher the intensity of business issues addressed by EA. Likewise, the higher the intensity of EA methods, tools and framework used by the organization, the higher the intensity of business issues that can be addressed by EA. In summary, the correlation analysis suggests that a more conducive environment for EA tends to address better business issues, whereas a stronger EA governance is likely to manage a wide-ranging business issues, and a more comprehensive methods, tools and framework to facilitate EA imlementation has a positive effect in addressing a wide coverage of business issues. Hence, the SEAM approach applied in this research is considered valid based on the empirical evidence presented in this section.

12.13 DISCUSSIONS AND CONCLUSIONS

In response to the first objective of the study, the main reasons for adopting EA found in this study are to support business and IT alignment, improve client's satisfaction and commitment, managing complexity, support systems development, support decision making, better work environment, and improve project management. Support business and IT alignment has been seen to be an important reason for EA activities in other prior studies. This suggests that organizations that want to ensure their business strategies are aligned to IT strategies should embark on EA. The significance of this alignment would ensure the organization's IT investment is justifiable. This is even strengthened by the second objective of this study, that provide evidence of the significance of the Systemic Enterprise Architecture Methodology (SEAM) as a viable approach in validating business-IT alignment.

EA is also considered important to manage road maps for change. Changes in enterprises are becoming fundamentally important because of the growing uncertainty in the global business environment today, therefore EA is important to manage these changes.

In terms of EA environment, the findings suggest that majority of the participating respondents acknowledged that their organizations are familiar with the importance of EA, though findings at the international level show a more overwhelming trend. This reveals that there is a growing interest in EA in the country, but actual EA adoption appears to be very minimal, particularly among the public sector organizations.

In terms of EA governance, the findings suggest that private sector organizations took greater efforts in making EA part of their strategic governance as compared to the public sector organizations. The findings also reveal that despite EA being considered important, it is largely the responsibility of IT managers instead of top management. This indicates a lower maturity index profile of EA governance. This is not surprising, since EA is considered a relatively new phenomenon in Malaysia. Studies done elsewhere (Schekkerman, 2005 and Matthee, 2007) show higher level of maturity with respect to EA implementations. These studies show a shift of responsibility for EA from IT managers to CEO and business managers.

On the extent of EA method, tools and framework used, dominant EA development methodologies are the Rational Unified Process and Rapid Application Development. Microsoft suites of tools almost sweep through the entire EA development activities. Majority also reported using house standards for EA framework, with growing interest in Service Oriented Architecture; whilst the more popular framework like the Zachman Framework was becoming less popular. This may suggest that organizations' preference depends much upon the industry and technology dominion, but at the same time organizations are caution in adopting new and emerging technologies. As a result, EA in Malaysia is slow to take off, but there is a growing interest among organizations in Malaysia towards EA as evidenced from this study.

With the SEAM approach proven viable, there is no excuse for organizations not to embark on Enterprise Architecture, as this study provides evidence that EA would be able to address an organization's business-IT alignment. Given the right environment with strategic governance in place and relevant methods, tools and framework for EA development, organizations would be able to achieve the returns on their IT investment and more importantly meeting their strategic business needs.

REFERENCES

1. Matthee, M. C., Tobin, P. K. J. & Van Den Merwe, P. (2007). "The Status Quo of Enterprise Architecture Implementation in South African Financial Services Companies," South Africa Journal of Business Management, vol. 38, pp. 11-23.
2. Patrick, P. (2005). "Impact of SOA on Enterprise Information Architectures," presented at SIGMOD, Baltimore Maryland USA.
3. Pereira, C. M. & Sousa, P. (2005). "Enterprise Architecture: Business and IT Alignment," presented at SAC "05, Santa Fe, NM, USA.
4. Rafidah, A. R., Zulkhairi, M. D., Huda, I., M-Khairudin, K. & Nor-Iadah, Y. (2009). "The Scenarios of Enterprise Architecture in Malaysian Organizations," presented at The 13th International Business and Information Management (IBIMA). Conference, Marrakech, Morocco.

5. Rafidah, A. R., Zulkhairi, M. D., Rohaya, D., Siti-Sakira, K., & Sahadah, A. (2007). "Enterprise Information Architecture (EIA).: Assessment of Current Practices in Malaysian Organizations," presented at Hawaii International Conference on System Sciences (HICSS-40)., Hawaii.

6. Schekkerman, J. (2005). 'Trends in Enterprise Architecture 2005,' Institute for Enterprise Architecture Developments (IFEAD).

7. Schekkerman, J. (2006). How to survive in the jungle of Enterprise Architecture Frameworks,Third Edition ed. Victoria, Canada: Trafford Publishing.

8. Seow, S. P. S. (2000). 'The Zachman Framework for Enterprise Architecture - Finding Out More,' USA: The Analyst LLC, 2000.

9. Watson, R. W. (2000). "An Enterprise Information Architecture: A Case Study for Decentralized Organizations," presented at The 33rd Hawaii International Conference on System Sciences, Hawaii.

10. Wegmann, A. (2003). "On the Systemic Enterprise Architecture Methodology (SEAM)," presented at the International Conference on Enterprise Information Systems, Angers, France.

11. Zulkhairi, M. D., Rafidah, A. R., Rohaya, D., Siti-Sakira, K. & Sahadah, A. (2006). 'Enterprise Information Architecture: An Assessment of Current Practices and Conditions,' Universiti Utara Malaysia, Sintok, Kedah, Research Report.

CHAPTER 13

ENTERPRISE ARCHITECTURE ONTOLOGY FOR SUPPLY CHAIN MAINTENANCE AND RESTORATION OF THE SIKORSKY'S UH-60 HELICOPTER

JAMES A. RODGER and PANKAJ PANKAJ

13.1 INTRODUCTION

This template, Enterprise Architecture is a comprehensive blueprint for the automated enterprise that is developed from different views and perspectives [1]. The thought was that a single company or organization can be an enterprise, when the reality is that no organizational entity can operate as a complete enterprise without considering 1) relationships with customers, 2) relationships with suppliers and contractors, and 3) relationships with regulators. Therefore, the enterprise model expressed in enterprise architecture must address how the operation interacts in its functional universe. Special consideration is given to 1) information, 2) processes, 3) enabling technology, and 4) enabling human resources and their associated organization.

This chapter was originally published under the Creative Commons License or equivalent. Rodger J and Pankaj P, Enterprise Architecture Ontology for Supply Chain Maintenance and Restoration of the Sikorsky's UH-60 Helicopter. Intelligent Information Management, *4,4 (2012), pp. 161-170. DOI: 10.4236/iim.2012.44024.*

Enterprise model or how an enterprise works as a collection of people, process, and technologies is often descriptive, ad hoc, or pre-scientific [2]. It is often a collection of heuristics which are not applicable in all circumstances. Enterprise architecture as a discipline has emerged to provide the theoretical foundations for designing and representing the enterprise in a scientific fashion. Using standard based ontologies to model enterprise is one of the core aspects of enterprise architecture.

Ontologies are of particular importance in the computer science and information systems area on account of their ability to model/represent knowledge as a set of concepts and the relationship between these concepts in a given domain. If an ontology is formulated in a crowdsourced/collaborative manner using a formal language then one can arrive at a formal unambiguous model/representation of the knowledge (referred to as ontology) about the given domain. Ontologies have developed for a variety of things like enterprise modeling [2], in-vivo biological cell types [3], marketing in relation to brand management [4], and socialism [5].

An enterprise is defined by a host of characteristics—processes, inputs, outputs, controls, enabling mechanisms including people (organizations) and technology. There are meaningful and dynamic relationships among these elements that affect cost and time as well as value and assets [1]. In the enterprise architecture domain an ontology (model) can account for all elements in the organization: its people, process, and technologies. The ontology can form the basis of common or shared understanding of the given domain can be used by internal constituents and partners of the enterprise for a variety of purposes like process integration etc. This shared conceptualization can also form the basis for design of information systems that can help run the enterprise more efficiently, and multi-enterprise collaboration using automated tools like intelligent agents. With regards to the latter Gubric and Fan provide an analysis of six supply chain ontologies [6]. Boeing's Boeing Technical Libraries developed a technical thesaurus in the form of a semantic network incorporating 37,000 concepts with an additional 19,000 synonym concept names, and 100,000 links [7-9] to promote common understanding between various partners involved in the manufacturing and design process.

It is well understood that that a modern enterprise must be data driven and all decisions should be based on information [1]. The process of preparing data for transformation into information and presenting it for action depends upon ontology. The process of associating information with experience, methods, and algorithms also depends on ontology. Given a problem domain within the context of an enterprise the first step would to represent the domain using an ontology. The ontology representation (syntax and semantics used to state the concepts and their relationships) should be based on standards especially if the domain spans across several enterprises. There are many standards available for representing an ontology [1] and one of the popular standard is the OWL 2 Web Ontology Language [10] by W3C.

In this paper we present a problem domain related to aircraft maintenance and provide description of the preliminary work done towards representing the problem domain by arriving at an ontology using the OWL ontology language.

13.2 MILITARY MAINTENANCE, REPAIR, AND OVERHAUL (MRO)

The military maintenance, repair, and overhaul (MRO) activities refer to the maintenance functions required to sustain an active aircraft fleet such as the Sikorsky UH- 60 [11]. The amount of maintenance required is directly related to the total number and usage of active aircraft. In other words, the greater the air time, the greater the maintenance demand, and the greater the MRO market. The MRO involves various constituents and complex relationships. Regulatory environment plays a key role in how the activities are carried out.

This is also an illustration of the complexity of the issues, and hence the requirement for an ontology that can promote shared understanding. MRO industry requires licenses from their suppliers. A key business segment for MRO firms involves obtaining PMA licenses. PMA licenses were enacted for two purposes. First, they monitor the quality of MRO replacement or modification parts for type-certified aircraft such as the Sikorsky UH-60. Second, they ensure a supply of MRO parts for all aircraft, both military and civilian. In a recent survey conducted by A.T. Kearney's Aerospace and Defense Practice [12], it was found that 96% of MRO respondents believed PMA parts to be among the top 10 issues facing the aerospace industry [13].

Sikorsky is the contractor and producer of the UH-60 helicopters series [12]. After each successful manufacturing, then the end product will be delivered to the military departments, who are its primary customers. However, each military department uses the helicopters for different types of missions and in different operating environments. Nevertheless, the DoD itself is able to exploit the Army facility to function as the central repair facility for all the UH-60 helicopter models. As mentioned before MRO is a complex operation involving many parties interacting in a complex fashion and therefore leads to several issues.

One of the key issues is that Sikorsky has little visibility on keeping track of each individual UH-60 record, because each military service records information differently in various flight data and maintenance log books [14]. The data is then captured as represented on forms. Each service has different form designs and records different data in terms of differences in methods and locations. Thus the Army and Sikorsky have difficulty in tracking the use and maintenance of the Navy and Army helicopters. When those helicopters arrive at the Army depot for repair, history and configuration management are investigated for security clearance. If Army does not have reliable data about those arriving helicopters at its depot, it will definitely create unnecessary and redundant replacements and exchange of parts, which will increase the operating costs and time.

In addition the Sikorsky's UH-60 is actually subject to the minimal time required constraint. In other words, adding more human resources and additional capital to the entire project will not lead to shorter MRO time. If more resources are added to the project, it will only increase the complexity and uncertainty of the whole project because more and more factors will be accumulated, and it is very hard to identify the underlying problems because groupthink phenomena will occur if there are redundant employees are hired for a project.

There are several ways to improve Sikorsky's UH-60 project and MRO. From a standards perspective, the government customer may adopt a top-down strategy and attempt to direct all of the services to adopt a single Information System. However, first, it has to recognize that for the Army, Navy, and Sikorsky to replace all of the disparate legacy systems is not possible, and it has to be noninvasive as possible. Alternatively, the solution would be a system of data exchange among all parties that assures accessibility to actionable data. By creating a data exchange mechanism like Enterprise Application Integration (EAI), an Enterprise Bus, or Service Oriented Architecture (SoA), a host or system neutral exchange mechanism can be provided. Such exchange mechanism would of course need unambiguous data definitions at the interface amongst other things. Sikorsky, Army, and Navy would be able to map the data required for exchange to any other qualified user in the aircraft community. Moreover, the standardization is a critical factor for the success of the whole integration process. In this case, one may use the ISO 13030, which is the Product Life Cycle Support Standard (PLCS) in defining and standardizing all the complications into one unifying and collaboration processing system. PLCS would alleviate the some discrepancies between the public sectors and the private sectors in transmitting the information.

Due to the different organizations involved in the MRO process and it complexity, it is important to arrive at the same shared conceptualization of the MRO domain between different participants. This conceptualization will be made conformant to the PLCS standard. As discussed before the domain can be represented with an ontology. This ontology can be constructed using a standards compliant language like OWL. The ontology can then be used to design information systems (for of data exchange and process standardization) across various partners involved in the Sikorsky UH-60 MRO to resolve the issues faced.

The following assumptions are made:

1. We have to assume that all the involved entities would be willing to use common models as a medium for data exchange that is applicable to most defense enterprise integration problems centered on exchanging information based on rigid standards and interfaces. For example, the Aircraft Maintenance Records entity from the ontology will have the data come from the Air Force Logistics Command (WR-ALC), Navy Air (NAVIR), Army Command (AMCOM), Suppliers, Depots, and Program Management Office (PMO).

2. The Army, the subcontractors, and the external suppliers are able to accommodate to the operating characteristics of the Navy

environment in providing the manufacturing, maintenance, and restorative services. Furthermore, we have to assume that Sikorsky will still maintain a different system for its internal operating environment, because it will provide some level of flexibility for Sikorsky in servicing those non-governmental contracts. Additionally, the military services are autonomous for its culture and it may have difficulty to embrace common processes and common needs.

3. The defense industrial business information objects can be completely converted into PLCS data exchange objects (DEX) without any errors. Besides that, there are some critical improvements in the data exchange between the dependent organizations. For instance, the data exchange capabilities will be elevated between Army and Sikorsky, Navy and Army, as well as Navy with Sikorsky. Moreover, we have to assume that we have 100% knowledge about the exact scope, product information, usage characteristics, and minimum requirements of the entire aircraft industries, which appropriately comply with the standards defined by the H60 Helicopter Program and Aviation Maintenance, the Federal Aviation Administration (FAA), and the Joint Aviation Authorities (JAA).

4. The Army, Navy and Sikorsky will adopt a standard and flexible data exchange utility, which enables seamless exchange without disrupting the respective application environments.

5. The data exchange processes are human-less processes, which mostly performed by autonomous computer system. There are no duplicated data or redundancy in database reporting, the responses to the request are almost spontaneous, and the involved data exchange utilities are flexible and reworkable at minimal efforts and costs.

6. This proposed aircraft MRO ontology is general and broad enough to cover the entire aircraft industry for Army, as well as Navy. This is needed for ensuring that the ontology is still useful and can be adapted to at least other similar aircrafts.

13.3 AIRCRAFT CLASS HIERARCHY

The ontology presented here is preliminary and a work in progress. In addition this work is primarily for research purposes and not oriented towards implementation. The complexity and scope of the problem precludes a complete presentation. It however provides a good illustration of the process of constructing an ontology using OWL Language and Protégé tool [15]. Protégé is a free, open source ontology editor and a knowledge acquisition system. Like Eclipse, Protégé is a framework for which various other projects suggest plugins. It is written in Java and heavily uses Swing to create the rather complex user interface. Protégé recently has over 160,000 registered users. Protégé is being developed at Stanford University in collaboration with the University of Manchester and is made available under the Mozilla Public License 1.1. The development of the aircraft ontology starts with the identification of the aircraft we identified for the case. Actually more aircraft types and classes can be added to this existing ontology we have developed as long as they shared common attributes, which can be linked to other entities. In our case, we only choose Boeing 777 (refer to Figure 1) and Sikorsky's UH-60 helicopter as our primary subjects for the development of the aircraft ontology. From the diagram below, we manipulated a variable, which is called Part_Type to connect the two different types of plane we investigated. It is because Sikorsky_UH-60 and Boeing_777 entities have recorded the information about the parts required for the maintenance purposes, which we will use in the later stage.

After we developed the major entities for the aircraft types, we will then further develop the sub-entities of Boeing 777 (refer to Figure 2) so that we will clearly observe the flow of the parts that are required to manufacture and to repair a Boeing 777 airplane. As we mentioned before, all the attributes (required parts and specifications) are linked to the Part_Type entity for the expansion of the entire ontology. The other aircraft type we have included for this aircraft ontology is the Sikorsky's UH-60 helicopter (refer to Figure 3). Similar to the Boeing 777 aircraft, we derive all the relative part levelby-level from Sikorsky_ UH-60 to the Tail_Section and the Front_Section, and then those two are further divided into sub-sub-parts or components. Similarly, all the attributes (required parts and specifications) are linked to the Part_Type entity for the expansion of the entire ontology.

13.4 OBJECT AND DATA PROPERTIES

We begin the process of capturing the domain knowledge pertinent to the repairable parts by focusing on the primary information flow. The primary objects in this partial ontology (refer to Figure 4) include air bases, air planes, types of parts, facilities and remote supply requests and depots. As we are required to understand the exact quantities of items such as parts and aircraft, but it is necessary to create a Quantity_Of class that permits the association of a numeric count with a specific plane or part type. In this way we can say that a particular facility has a Quantity_Of instance relating a particular item with a specific number. It was also necessary to be able to associate each air base with an ordered list of remote supply facilities available to provide additional parts, which can be achieved using an Remote_Supply_Requests structure (will be connected to other variables later in the complete ontology).

Simulated data was constructed for this scenario consisting of the inventory of aircraft and parts at from different air bases and different remote supply bases taken at various times. Each event contained facility-specific information such as the quantity of good aircraft of each type, the quantity of aircraft parts in stock, and the quantity of fixable parts in stock along with the current need for parts that needed to be replaced on aircraft undergoing repair. In addition to this event data, a file of annotations was created containing descriptions of the various aircraft types and the parts that make them up, while another annotation file was constructed to provide descriptions of the specific air bases, their aircrafts and their remote supply facilities. In other words, the constructed entity relationship diagrams above will illustrate the six interrogatives of what, when, who, where, why and how, described by the Zachman Framework [16].

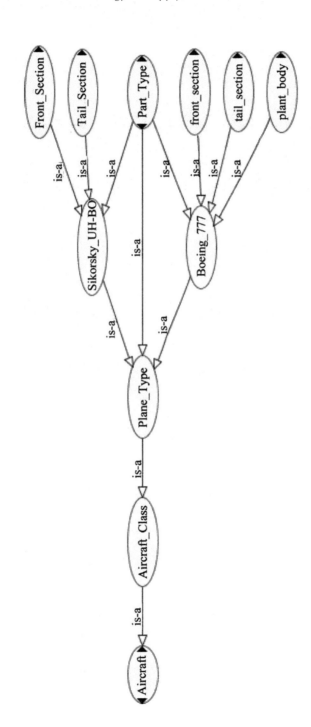

FIGURE 1: Aircraft class entity relationship diagram

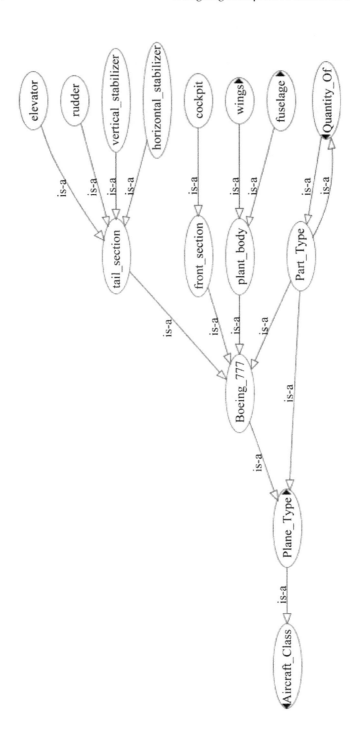

FIGURE 2: Boeing 777 entity relationship diagram

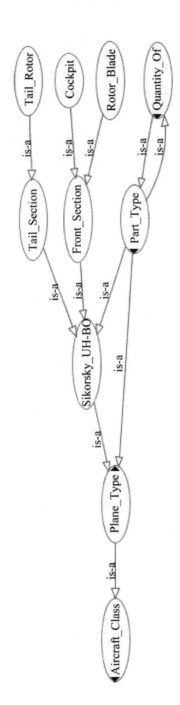

FIGURE 3: Sikorsky's UH-60 helicopter entity relationship diagram

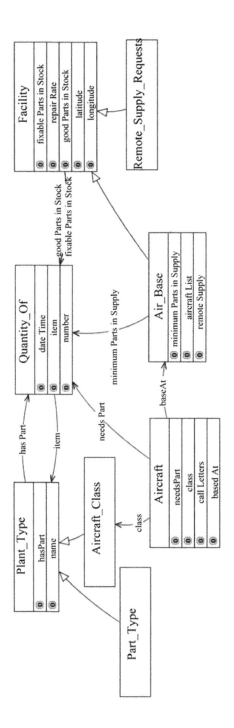

FIGURE 4: Partial aircraft entity relationship diagram and its attributes

After understanding how the information flow between the internal aspects of the Army's, Navy's, and Sikorsky's information systems, the next step we have to configure the external working environments for the maintenance requests to be delivered to the suppliers and the appointed subcontractors. First, we will implement a SCM system, which acts as a middleware for information exchange between the buyer side and the seller side. It pretty much links all the variables such as Facility, Air_ Base, Remote_Supply_Requests, Aircraft_Maintenance_ Records, SCM_Data_ Repository, Distributors, and PLCS (will discuss in later section).

The topology of this SCM is defined by entities and connectors (refer to Figure 5). The entities interact with each other through the connectors as an order is fulfilled in a supply chain. Each entity performs five main actions with regard to the order life cycle, which are the creation, placement, processing, shipping, and finally receiving. Those orders are initiated based on the Remote_Supply_ Requests, placed by the Army, or the private sectors. The order transport is to be assumed with some level of processing delays. Once the requests are initiated, the order will be shipped from the supplier side to the initiation entity. Whenever the orders are received, they will be consummated immediately (no inventory for stock items).

We basically have identified the primary supply chain agents in the SCM system, and they are distributors, assemblers, Manufacturers, and suppliers (refer to Figure 6). Each connector between those entities serves as the tracking and coordination utility for the flow of material, information, and finance in the supplier-customer network.

We will further develop the supplier network by identifying all the participating suppliers, which will contribute to the manufacturing and maintenance of the aircrafts (refer to Figure 7). In our case, we develop an interactive supplier database system for all the participating suppliers for the bidding activities to take place. Suppliers who comply with the PLCS specifications will be chosen to be the prime contractors for the Army, Navy, and Sikorsky.

Next, we have to impose the Product Life Cycle Support Standard (PLCS), ISO 10303, to map the order specifications against the aircraft maintenance records by the military users (refer to Figure 8). Thus, PLCS serves as the catching mechanism to filter the data redundancy and faulty parts, which are not appropriately, comply with the PLCS specifications.

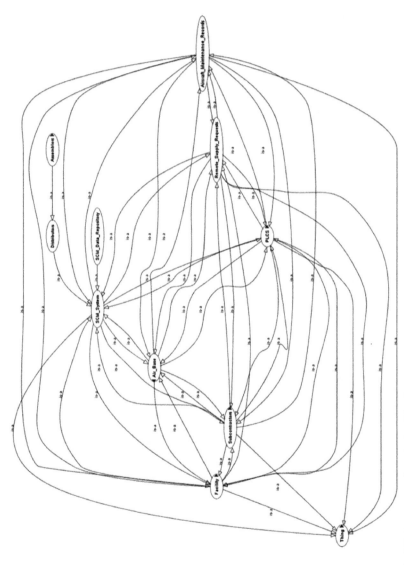

FIGURE 5: SCM system entity relationship diagram.

FIGURE 6: Simplified SCM entity relationship diagram

PLCS entity contains the PLCS_Data_Repository, which consists on all kinds of forms for internal processing purposes. It is needed for data exchange, visibility, and flexibility.

In addition to that, we have to consider that sometimes Army from the air bases and facilities may employ outside contractors to perform the maintenance activities, when the Army side lacks of latest knowledge and expertise. Once again, those subcontractors will be interacted with the SCM systems, to comply with the PLCS standards (refer to Figure 9), and to follow the Aircraft_ Maintenance_Records specifications from various air bases, different facilities, and at different time. Thus, we have to establish interrelated connectors between those entities.

As we connect all the partial ontologies for the maintenance of the aircraft by the Army for the Navy people, we will be able to understand the overall picture of the complete aircraft ontology (refer to Figure 10). First, we will realize that all the significant entities are actually interrelated, interoperated, and dependent in nature. Second, as we discover more factors, we have to actually increase the entities in the ontology, which in turn, increase the complexity of the ontology. Moreover, we will be able to observe the critical path factors in the aircraft ontology that we have created, from the number of connectors that an entity has.

13.5 CONCLUSIONS AND FUTURE WORK

An ontology can present the knowledge about a domain in a scientific and unambiguous fashion. This representation can be used for a common shared understanding by different constituents of the domain including various stakeholders and can also be used as the basis for designing information systems. Here we have presented an ontology that modifies the current domain to be standard compliant and therefore provides a solution that may alleviate some issues through design of appropriate information

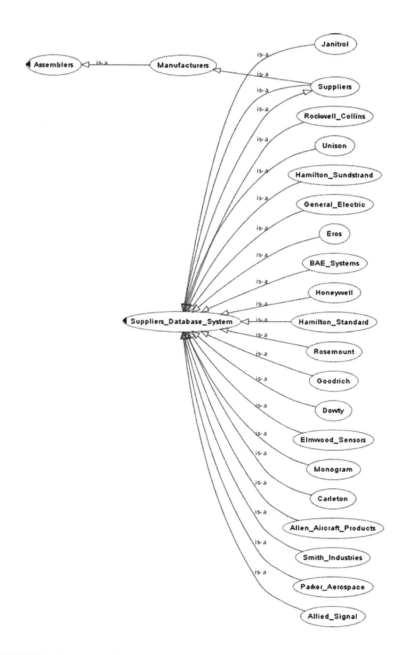

FIGURE 7: Supplier database system entity relationship diagram.

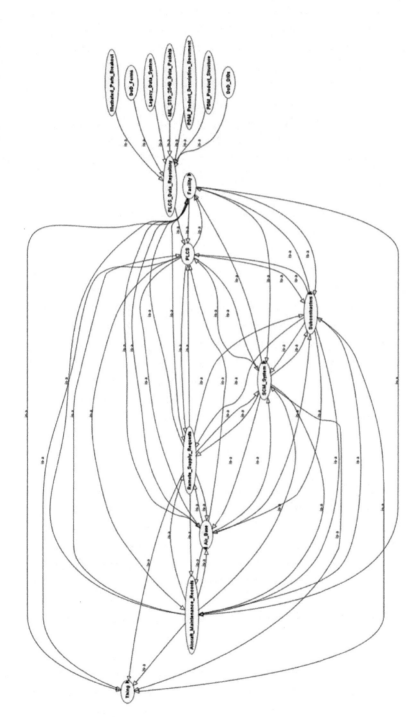

FIGURE 8: PLCS entity relationship diagram.

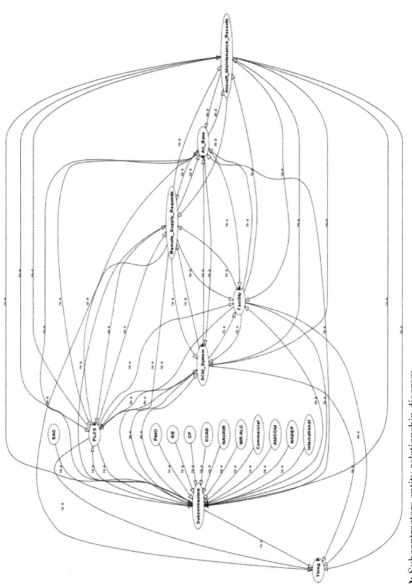

FIGURE 9: Subcontractors entity relationship diagram.

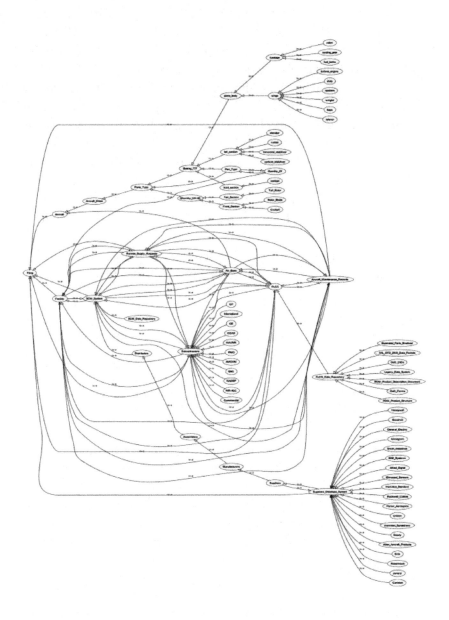

FIGURE 10: Complete Boeing 777 and Sikorsky's UH-60 helicopter ontology.

systems. With reference to the example issue that was discussed model-driven data exchange may be implemented as a replacement for point-to-point and hub-spoken connectivity. Additionally international standard framework for data exchange may be used so that open interoperability may be achieved and entire ontology can be expanded.

The ontology presentation in this paper is somewhat limited fashion due to issues of scope and functionality. The ontology is preliminary based on available information and may be improved. To improve the proposed ontology, more recent data would be gathered so that some unsolved exceptions can be handled effectively. Additional development spirals for the ontology and associated information systems, could be undertaken from time to time through gathering of more performance data and learned-from-experienced data from the customer satisfaction reports. Next one may focus on the efficiency and improvement of planning, problem solving, sense making, and the decision making of the maintenance activities and through timely and reliable forecasts on the arrival of those aircrafts to the maintenance facilities.

REFERENCES

1. J. J. George and J. A. Rodger, "Smart Data: Enterprise Performance Optimization Strategy," John Wiley & Sons, Inc., Hoboken, 2010.
2. M. Gruninger and K. Atefi, "Ontologies to Support Process Integration in Enterprise Engineering," Computational & Mathematical Organization Theory, Vol. 6, No. 4, 2000, pp. 381-394. doi:10.1023/A:1009610430261
3. T. F. Meehan and A. M. Masci, "Logical Development of the Cell Ontology," BMC Bioinformatics, Vol. 12, No. 1, 2011, p. 6. doi:10.1186/1471-2105-12-6
4. W. Grassl, "The Reality of Brands: Towards an Ontology of Marketing," The American Journal of Economics and Sociology, Vol. 58, No. 2, 1999, pp. 313-359.
5. R. Westra, "Marxian Economic Theory and an Ontology of Socialism: A Japanese Intervention," Capital & Class, Vol. 26, No. 3, 2002, pp. 61-85. doi:10.1177/030981680207800104
6. T. Grubic and I. S. Fan, "Supply Chain Ontology: Review, Analysis and Synthesis," Computers in Industry, Vol. 61, No. 8, 2010, pp. 776-786. doi:10.1016/j.compind.2010.05.006
7. P. J. Clark and R. Thompson, "Exploiting a Thesaurus- Based Semantic Net for Knowledge-Based Search," 12th Conference on Innovative Applications of AI (AAAI/ IAAI'00), 2000. http://www.aaai.org/Papers/IAAI/2000/IAAI00-008.pdf
8. A. Hunter, "Engineering Ontologies," 2001. http://www.cs.ucl.ac.uk/staff/a.hunter/tradepress/eng.html

9. M. Uschold, "Creating, Integrating and Maintaining Local and Global Ontologies," First Workshop on Ontology Learning (OL-2000) in Conjunction with the 14th European Conference on Artificial Intelligence (ECAI 2000), Berlin, 2001.

10. P. Hitzler and M. Krötzsch, "OWL 2 Web Ontology Language Primer," 2012. http://www.w3.org/TR/owl2-primer/

11. L. Miller and M. Bertus, "License Valuation in the Aerospace Industry: A Real Options Approach," Review of Financial Economics, Vol. 14, No. 3-4, 2005, pp. 225-239. doi:10.1016/j.rfe.2005.04.001

12. Wikipedia, "Sikorsky UH-6," 2012. http://en.wikipedia.org/wiki/Sikorsky_UH-60

13. W. Anderson, "Logistics and Support Chain Management: An Aerospace Industry Perspective," 5th Annual Executives Logistics Forum, University of North Texas, Denton, 2001

14. C. Koblish, "MRO—Crisis or Cycle," Maintenance, Repair, and Overhaul Conference Fort Lauderdale, McGraw Hill, New York, 2003.

15. Standford, "Protégé," 2012. Protege.standford.edu

16. J. A. Zachman, "John Zachman's Concise Definition of the Zachman's Framework," 2008. http://www.zachman.com/about-the-zachman-framework

CHAPTER 14

AN ENTERPRISE ARCHITECTURE FRAMEWORK FOR MOBILE COMMERCE

KHAWAR HAMEED, HANIFA SHAH, KAMRAN AHSAN, and WEIJUN YANG

14.1 INTRODUCTION

The phenomenon of mobility has driven an evolution and revolution of technologies and new ways of working, exploiting the practice of anytime, anyplace, anyhow computing whilst gaining critical mass as a research discipline and paradigm - the study of which has provided a grounding for conceptual and theoretical perspectives that allow framing and discussion of mobility [2] [3]. Mobile commerce has created a wide range of business opportunities—the spectrum of which includes the transformation of mobile network operators to go beyond the provision of core voice services to the delivery of value added services and service aggregation that provide comprehensive consumer packages [4], the emergence of business models to develop and deploy mobile applications (such as the Apple 'App' Store, the O_2 Litmus programme, and the Sony Ericsson Developer World – most of which are based on shared revenue distribution through sales through the respective hosting channels. Furthermore,

*This chapter was originally published under the Creative Commons License or equivalent. Hameed K, Shah H, Ahsan K, and Yang W. An Enterprise Architecture Framework for Mobile Commerce. Interna-*tional Journal of Computer Science Issues, *7,3 (2010), ISSN (Online): 1694-0784.*

open programmes and platforms such as Android (Google) are likely to contribute to the enlargement of the developer community that seeds the growth of mobile applications. With the supply chain of mobile computing and commerce, a range of opportunities exist for contributing parties to collaboration to provide value-added solutions based on new relationship types, rules and ecosystems [5] further enhancing the composite capability and reach of service providers into new market areas. Beyond service providers, the ease of access, penetration, and diffusion of mobile technologies has enabled individual sectors and organisations therein to apply the concepts and practice of mobility to create innovative domain-specific m-commerce applications that leverage the specific nuances associated with those domains resulting in value-added solutions for end-users and new revenue-generating opportunities for business. Examples of these include mobile location-based tourism, travel and navigation systems, m-ticketing and booking applications. Beyond the core commercial imperative of increased revenue generation, the application of mobile computing and principles of m-commerce also underpin the development of private and public sector mobile applications that aim to reduce operating costs, enhance efficiency and provide better platforms for engaging the end-user population. Basole [2] presents an Enterprise Mobility Continuum that frames mobile solutions from point-specific solutions to those that diffuse across entire organisations to create stakeholder value. Across and throughout the complex m-commerce ecosystem the question arises regarding the construction of m-commerce solutions and how these are best approached.

The increasing attention to business underpinned by mobility, mobile services, mobile applications, and technologies has become a major driver for the development of m-commerce systems. This growing trend has become a focus for a significant number of organisations. This paper proposes that in developing mcommerce systems organisations need to establish an enterprise architecture for m-commerce. The rationale for this is rooted in the need to develop a holistic and integrated view of strategic direction relating to mcommerce which will enable a coordinated and controlled approach that reduces complexity and yields effective systems based on the structured integration of services, practices and technology resources. In doing so, the potentially complex universe of discourse associated with mobility is harnessed to produce and organisational asset to drive the development of m-commerce.

The next section proposes a framework to establish such an enterprise architecture for mobile commerce. Firstly, an enterprise architecture framework for mobile commerce is presented followed by a brief view of associated issues concerned with method.

14.2 AN ENTERPRISE ARCHITECTURE (EA) FRAMEWORK FOR MOBILE COMMERCE

An EA framework provides the basis or template for the creation and establishment of enterprise architecture. Zachman [6] is credited with developing the discipline of enterprise architectures as a concern for both researchers and practitioners. An enterprise architecture framework is essentially a meta-construct used to define the scope of the associated architecture and how the areas of the architecture relate to each other [7]. An architecture can be considered analogous to a blueprint or plan of a building structure, where different perspectives may exist and each perspective contains structures that demonstrate inter-relationships based upon some predefined constraint and yield a solid foundation and approach upon which the building is constructed.

Generally speaking, the EA framework defines the scope of the resulting architecture, which typically includes a business view, information integration, application-level views, and technology infrastructures. Definitions of each view may include more refined constructs and relationships at a lower level of granularity.

The application of an EA approach is considered relevant and appropriate since the ecosystem within which the development of m-commerce solutions occurs comprises a set of interrelated perspectives based upon the integration of mobile devices, technologies, and business processes [8]. Therefore, an EA framework for mobile commerce can be considered to address, at least, the scope of architecture covering the business level/view, the application level and the technology infrastructure level.

In each level of our proposal, core components have been identified. These are: a business model of m-commerce (business level), supply chain of m-commerce (supply chain level), m-commerce applications (application level) and technology infrastructure for m-commerce (technology level). Our research has shown that integration of core EA approaches with m-commerce is relatively sparse, whereas literature in both contributing areas is substantial. Furthermore, Leist & Zellner [9] state that "as information systems grown in complexity and scope the need for a comprehensive and consistent approach in modelling these systems becomes of paramount importance." Basole [2] recognises that businesses are "just beginning to recognise the importance and potentially transformative impact of enterprise mobility." Given this our approach to applying EA principles to m-commerce appears well-grounded. A proposed Enterprise Architecture Framework for mobile commerce is shown in Figure 1.

14.2.1 FIRST LEVEL: M-COMMERCE BUSINESS MODELS

Figure 1 shows the first (topmost) level in our Enterprise Architecture framework for m-commerce relating to business models. This level is identified as the first level in the EA framework and provides a description of the roles and relationships of an organisation, its customer, partners and suppliers and stakeholders, as well as the flows of goods, information and money between these parties and the main benefits for those involved [10]. The stakeholder transactional models are based upon those presented by Coursaris & Hassanein [11] and are divided into four models (i) wireless Business-to-Consumer (W_{B2C}) model; (ii) a wireless Consumer-to-Business (W_{C2B}) model; (iii) a wireless Consumer-to-Consumer (W_{C2C}) model; and (iv) a wireless Consumer-to-Self (W_C^2) model. These models mainly describe business activities between the contributing parties, and the specific nature of these. The abstraction of these models provides a basis for higherlevel understanding of the spectrum of stakeholders and relationships at the topmost level of the architecture.

Other business models of mobile commerce attempt to address the complexities associated with the m-commerce ecosystem comprising different participants (such as mobile consumers, network operators, service

FIGURE 1: An Enterprise Architecture Framework for Mobile Commerce

providers, application developers, content providers and technology providers) and encompass the types of services and sources of profits. In addition, models describe operations and processes relating to mobile growth, value-added benefits, revenue models and return on investment, and the transfer of benefits across stakeholders for mutual gain [12].

During the development of our framework we reviewed models broadly associated with the business models for mobile information, mobile advertising and for mobile office work. Typical characteristics of these were associated with relationships between content providers, network operators and service providers for direct gain through revenue opportunities from mobile consumers and shared distribution amongst stakeholders and participants. Interestingly, one specific finding was that, "mobile consumers represent the only sustainable revenue source for participants" [13].

From the characteristics identified, it is seen that business models for mobile commerce are helpful in enabling organisations to increase the possibilities for increasing revenue and profit, and enhancing competitiveness. Taking Vodafone Group Plc [14] as a typical example rather complex business activities can be observed where the business acts as wireless network operator providing information transmission services for its customers whilst also acting as a content or service providers of specific service content for its customers such as music downloading, online gaming, e-mail and location-based information. In this and other similar cases the business models appear not to be mutually exclusive (e.g. single and independent W_{B2C}, W_{C2B}, W_{C2C}, W_C^2 business models) [11]. Thus, encapsulating the diverse stakeholder groups and associated mobile information needs can be seen as providing competitive advantage and market diversification.

14.2.2 SECOND LEVEL: THE SUPPLY CHAIN OF MOBILE COMMERCE

Figure 1 shows the second level in the Enterprise Architecture framework for mobile commerce and is concerned with the supply chain. The materialisation of business models for m-commerce depends on a complex chain of business relationships between participants of the supply chain [12]. The supply chain for m-commerce is therefore identified as the second and supporting level in the enterprise architecture framework.

This supply chain for m-commerce can be seen as rooted in mobile telecom markets, within which a variety of participants possess resources, perform activities, and are in relationships that are established or evolving in delivering an end-to-end service [15]. However, given the increasing diversity of applications, services and associated information delivery, this supply chain is being extended to include a spectrum of market areas which, when aggregated provide value-added services to the mobile user.

The core elements of the m-commerce supply chain include mobile commerce participants, mobile commerce resources and mobile commerce activities. Mobile commerce participants are a major driving force behind the m-commerce supply chain [13]. Mobile commerce participants fundamentally include mobile consumers, wireless network providers, content provide, service providers, application developers and technology providers—all of whom engage in a business-oriented relationship and, in the specific case of m-commerce, one that is focussed on commercial gain for service delivery participants. For example, the shared/distributed revenue approach entails the wireless network providers delivering part of that revenue to other participants in the supply chain—such as content providers and other service providers.

14.2.3 THIRD LEVEL: MOBILE COMMERCE APPLICATIONS

The third level of the framework is concerned with mobile commerce applications. Because an organisation uses mobile applications to support and deliver its business models through the supply chain, mobile applications are seen as the tangible end-user vehicles that mechanise and enable the m-commerce transaction. Mobile applications are therefore identified at the third level in the Enterprise Architecture framework for m-commerce.

M-commerce applications can be broadly categorised as communication applications, information applications, entertainment applications and commerce applications [11]. Five main application types were identified: mobile ticketing, mobile advertising, mobile information, mobile banking and mobile office applications. These were found to support the key business models and imperatives (revenue generation or cost reduction) and were constructed using the supply chain to deliver an aggregated service. The constantly developing landscape of mobile technologies, and more specifically application capability, raises a proposition of re-aligning business models and supply chains to fully leverage the potential of that change—thus suggesting a commercially synergistic relationship (in this case, the applications and technology forcing a re-evaluation of the business models and supply chain) .

14.2.4 FOURTH LEVEL: THE TECHNOLOGY INFRASTRUCTURE FOR MOBILE COMMERCE

The fourth level of the framework is concerned with technology infrastructure for m-commerce. This includes wireless communication technology, wireless middleware technology, information exchange technology, wireless network & application protocols and mobile security technology. Essentially, these are the core technological components and infrastructures that enable mobile users in their environments. These technologies support upper levels of the framework (m-commerce business models, m-commerce supply chain, and m-commerce applications).

14.3 METHOD

In order to materialise a specific architecture from the enterprise architecture framework an associated method is required. The method provides a step-by-step description of how to establish the architecture [7]. In context of the proposed framework, a five-step approach is proposed based upon five generic steps [16] adapted to suit the framework-specific needs. Our current research focuses on the development of the m-commerce enterprise architecture framework and the method associated with this is in embryonic form. Nevertheless, views and expectations are that this will focus on internal knowledge audits to determine organisational readiness and mobile strategies, a series of data collection methods that aim to elicit the type of existing of new applications that can be mobilised in context of m-commerce, the skills base, and the technology/resource levels within the organisation, and levels of innovation. Other aspects are expected to address appropriate business planning tools, modelling techniques and notations that apply across the four levels of the framework and incorporate the modelling of mobility in context of mobile commerce opportunities. Technical limitations of mobile devices and wireless communications, business concerns and legal constraints complicate the practical use of mobile commerce [17]. Therefore, these also provide the impetus of developing appropriate methods that address these concerns. In all these cases organisations may have existing approaches and techniques that might be used a part of the method.

14.4 VALIDATION

To validate the proposed framework a questionnaire was developed to test the overall construct of the framework, its direct relevance to businesses, and to seek the views that might be incorporated in its refinement. The survey aimed to elicit initial and relatively informal feedback and as such was constructed to provide qualitative feedback that would enable scope for interpretation and discussion. Three organisations (anonymity maintained) participated in the survey. Questions were asked in two parts (EA Framework for M-Commerce & EA Method for MCommerce) and sought views on the following areas and propositions:

14.4.1 THE BUSINESS MODEL FOR M-COMMERCE

- A1: The positioning of the business model for mcommerce at the first (topmost) level in the EA Framework
- A2: The role of a business model for m-commerce to describe how organisations create and realise genuine benefit from mobile commerce activities.
- A3: The division of an m-commerce business model into a Wireless Business-to-Consumer model, Wireless Consumer-to-Business model, Wireless Consumer-to-Consumer model, and Wireless Consumer-to-Self model [11].
- A4: The composition of a business model for mcommerce to include the source of profits, mobile services and supply chain participants.
- A5: The choice of novel or more recent business model approaches over traditional business models for m-commerce.

14.4.2 THE SUPPLY CHAIN FOR M-COMMERCE

- A6: The positioning of the supply chain model for mcommerce at the second level in the EA Framework.

- A7: That a good supply chain for m-commerce can streamline business processes for m-commerce.
- A8: That the supply chain for m-commerce includes m-commerce participants, m-commerce resources and m-commerce activities.
- A9: That the m-commerce participants are a major driving force behind the supply chain of mcommerce.

14.4.3 M-COMMERCE APPLICATIONS

- A10: The positioning of m-commerce applications at the third level in the EA Framework.
- A11: M-commerce applications are influenced by the business model and supply chain for m-commerce.
- A12: M-commerce applications need be supported by reliable m-commerce technology

14.4.4 TECHNOLOGY INFRASTRUCTURE FOR M-COMMERCE

- A13: The positioning of technology infrastructure for m-commerce at the final level (bottom) in the EA Framework.
- A14: The role of mobile communication technologies in offering message or wireless data capability.
- A15: That wireless middleware technology supports the development and operation of m-commerce applications.
- A16: Information exchange technologies are necessary for m-commerce applications and mobile devices.
- A17: That wireless network and application protocols can support the delivery of web-based information.

14.4.5 EA METHOD FOR M-COMMERCE

- B1: The positioning of data collection (knowledge audit) as a first step in an EA method for m-commerce.
- B2: Data collection should comprise identification of m-commerce components and also possible obstacles or restrictions in m-commerce.
- B3: That defining the purposes of an EA for mcommerce is defined as the second step.
- B4: That establishing an EA (schema) for mobile commerce is defined as the third step.

- B5: That an EA framework of m-commerce needs be modelled by software tools.
- B6: That tool selection is defined as the finall step.

14.4 RESULTS

A 3 scale scoring system was used and responses constrained to Accept (3), Neither/Neutral (2), and Not Accepted (1). All three participants responded and the summarised results for each participant are tabulated as shown.

TABLE 1: The total feedback score: Organisation 1

Establishing Enterprise Architecture (EA) for Mobile Commerce (M-Commerce)	
Section A: An EA Framework of M-Commerce	
The Business Model of M-Commerce:	Total scale: (1–15) A1: (3) + A2: (3) + A3: (2) + A4: (3) + A5: (3) = 14
The Supply Chain of M-Commerce:	Total scale: (1–12) A6: (3) +A7: (3) + A8: (3) + A9: (2) = 11
M-Commerce Applications:	Total scale: (1–9) A10: (3) + A11: (2) + A12: (3) = 8
Technology Infrastructure of M-Commerce:	Total scale: (1–15) A13: (3) + A14: (3) + A15: (3) + A16: (3) + A17: (1) = 13
Section B: An EA Method for M-Commerce	
The Steps in an EA Method for M-Commerce	Total scale: (1–18) B1: (3) + B2: (3) + B3: (2) +B4: (3) + B5: (3) + B6: (2) = 16

TABLE 2: The total feedback score: Organisation 2

Establishing Enterprise Architecture (EA) for Mobile Commerce (M-Commerce)	
Section A: An EA Framework of M-Commerce	
The Business Model of M-Commerce:	Total scale: (1–15) A1: (3) + A2: (3) + A3: (2) + A4: (3) + A5: (3) = 13
The Supply Chain of M-Commerce:	Total scale: (1–12) A6: (3) +A7: (3) + A8: (3) + A9: (2) = 11
M-Commerce Applications:	Total scale: (1–9) A10: (3) + A11: (2) + A12: (3) = 8
Technology Infrastructure of M-Commerce:	Total scale: (1–15) A13: (3) + A14: (3) + A15: (3) + A16: (3) + A17: (2) = 14
Section B: An EA Method for M-Commerce	
The Steps in an EA Method for M-Commerce	Total scale: (1–18) B1: (3) + B2: (3) + B3: (2) +B4: (3) + B5: (3) + B6: (3) = 17

TABLE 3: The total feedback score: Organisation 3

Establishing Enterprise Architecture (EA) for Mobile Commerce (M-Commerce)	
Section A: An EA Framework of M-Commerce	
The Business Model of M-Commerce:	Total scale: (1–15) A1: (3) + A2: (2) + A3: (2) + A4: (3) + A5: (3) = 13
The Supply Chain of M-Commerce:	Total scale: (1–12) A6: (3) +A7: (2) + A8: (3) + A9: (2) = 10
M-Commerce Applications:	Total scale: (1–9) A10: (3) + A11: (3) + A12: (2) = 8
Technology Infrastructure of M-Commerce:	Total scale: (1–15) A13: (3) + A14: (3) + A15: (3) + A16: (3) + A17: (2) = 14
Section B: An EA Method for M-Commerce	
The Steps in an EA Method for M-Commerce	Total scale: (1–18) B1: (3) + B2: (3) + B3: (2) +B4: (2) + B5: (3) + B6: (2) = 15

All 3 respondents scored highly regarding the business model component of the proposed framework, indicating its appropriate and relevant positioning. Similarly, responses regarding the supply chain model component of the framework scored highly. This again was interpreted as a positive response regarding its overall context and positioning. The scoring for m-commerce applications as part of the framework was slightly lower, although this was still deemed a positive response. Finally, the technology infrastructure level scored highly, again indicating an overall positive response. The overall interpretation was that the framework proposition was an appropriate and useful vehicle for the development of mcommerce within organisations.

With respect to the method, all responded again scored this aspect highly thus indicating its potential relevance, value and use.

14.5 CONCLUSIONS

Mobile technology is continuing to play a significant role in providing efficient and effective means for organisations to broaden their revenue streams, and enhancing competitive positioning. There is no doubt that the mobile ecosystem creates many possibilities for the development of innovative solutions that create real value for end users whilst meeting real financial imperatives of service providers and stakeholders in the m-commerce supply chain. However, the mobile ecosystem is potentially complex and the universe of discourse created contains many interrelated components that span across financial, organisational, technological, and social boundaries. It is this that provides the impetus and rationale to focus on adopting an enterprise architecture framework approach to govern the development of mcommerce systems and with a view to reducing complexity. It is also this that aims to enable organisations to adapt rapidly to mobile technologies, reap the affordances created through technology adoption, remain competitive—yet through a structured and framework-oriented approach that provides a baseline for pro-active transformation rather than a re-active and potentially chaotic and fragmented approach.

This paper has proposed an initial enterprise architecture framework for mobile commerce that aims to provide practitioners and researchers a platform for considering the development of m-commerce systems, and one which aims to influence both philosophical and practical approaches to building m-commerce systems. Initial response, albeit being based on limited demographics, has been positive and this provides motivation for future work. The framework proposition comprises four levels, each of which draws upon specific nuances associated with the mobile environment and each of which is constructed in context of other layers—thus demonstrating a frameworkoriented approach.

REFERENCES

1. Mahatanankoon, P., Wen H.J., & Lim,B. (2005) Consumer-Based m-commerce: Exploring Consumer Perception of Mobile Applications. Computer Standards & Interfaces, 27(4), 347-357.
2. Basole, Rahul C., Enterprise Mobility: Researching a New Paradigm. Information Knowledge Systems Management 7 (2008) 1-7.
3. Kakihara, M., Sorensen, C. (2001) Expanding the 'Mobility' Concept. ACM SIG-GROUP Bulletin. 22 (3) 33-37.
4. Seybold, Andrew M., The Convergence of Wireless, Mobility, and the Internet and its Relevance to Enterprises. Information Knowledge Systems Management 7 (2008) 11-23.
5. McDowell, M., Business Mobility: A Changing Ecosystem. Information Knowledge Systems Management 7 (2008) 25-37.
6. Zachman, J.A., A Framework for Information Systems Architecture. IBM Systems Journal, Vol 26 (3) 1987.

7. Scott A. Bernard, An Introduction to Enterprise Architecture, Second Edition, 2006.
8. N.Bandyopadhyay, E-Commerce: Context, Concepts and Consequences, Published by McGraw-Hill Education, Second Edition, 2002.
9. Zellner, S. L. A. G. (2006). Evaluation of current architecture frameworks. ACM, SAC'06, April, 23-27, 2006, Dijon, France.
10. Giovanni Camponovo and Yves Pigneur, Business Model Analysis Applied To Mobile Business, The University of Lausanne, 2004.
11. Coursaris & Hassanein, Understanding m-commerce: A Consumer-centric Model, Degroote School of Business, McMaster University, 2006.
12. WANG Yan & GAO Yufei, Research on the Value Chain and Business Models for Mobile Commerce, Fudan University, 2004.
13. Brian E.Mennecke and Troy J.Strader, Mobile Commerce: Technology, Theory, and Applications, Lowa State University, Published by Idea GroupInc, 2003.
14. Vodafone, About Vodafone Power to You, from http://www.vodafone.com/start/about_vodafone, accessed November 13, 2009.
15. Nan Si Shi, Wireless Communications and Mobile Commerce, Published in the United Kingdom by Idea Group Inc, 2004.
16. Rob.C.Thomas, A Practical Guide for Developing an Enterprise Architecture, Federal Chief Information Officer Council, 2004.
17. Keng Siau, Ee-Peng Lim, Zixing Shen, Mobile Commerce: Current States and Future Trends, University of Nebraska-Lincoln, USA, 2006.

CHAPTER 15

EA-MDA MODEL TO RESOLVE IS CHARACTERISTIC PROBLEMS IN EDUCATIONAL INSTITUTIONS

MARDIANA and KEIJIRO ARAKI

15.1 INTRODUCTION

Educational institutions play a pivotal role in society, primarily in developing countries. Educational institutions which in this case refer to higher education institutions (HEI) have business domains unlike that of business entities or other organizations. HEI or university places more emphasis on the role of information technology (IT) in supporting its business processes. Given its scale and complexity, IT management in HEI can be categorized as enterprise-scale data management. As such, particular planning and design is necessary to ensure that IT is applied in accordance with the institution's strategic objective and plan, and that it can be optimally utilized by users. The purpose of HEI in Indonesia adheres to the Triple Principles of Higher Education Institutions (Tridharma Perguruan Tinggi) that comprise education, research and community service. To ensure adherence to these principles, every university has developed their own strategic plan according to their vision and mission statements. University must also synchronize their business and technology strategies. All of these can be attained by applying the enterprise architecture (EA)

This chapter was originally published under the Creative Commons License or equivalent. Mardiana and Araki K. EA-MDA Model to Resolve IS Characteristic Problems in Educational Institutions. International Journal of Software Engineering & Applications *4,3 (2013), DOI : 10.5121/ijsea.2013.4301.*

approach within the institution. Through the optimal performance of business processes, EA can facilitate strategic planning, align business and IT resources, and regulate the lifecycle of an information system (IS) development process [1].

IS development is an aspect that should be part of an institution in order to support business activities and provide services to stakeholders, mainly in relation to data, information, technology and application. IS development should be well planned, centralized or distributed in related working units, and integrated into other existing systems. IS has helped automate many phases in the business process previously done manually in an educational institution. In line with shifts in user needs, changes to the business process are unavoidable. If an information system fails to accommodate these changing trends, its utilization will therefore be less than optimal. Hence, such changes need to be taken into account when developing an IS.

Service-oriented Architecture (SOA) helps accommodate changing business needs by providing flexibility for more efficient and effective use of IT resources. Through SOA, substantial and complex business processes are broken down into smaller and simpler services that allow for easier and faster changes to business processes [2]. SOA is equally useful for dealing with issues related to the integration of various existing systems, maintenance and improving application and systems performance. SOA however requires platform-independent services, including services that can translate all business service needs into different implementations of information technology. Model-Driven Architecture (MDA) is capable of addressing SOA weaknesses. MDA helps create an model that IS developers can make use of, which ranges from goals and requirements specification to implementation through several abstraction levels [3].

Applying the EA, MDA and SOA models together is the key element for optimizing the business processes of an enterprise. EA provides a comprehensive understanding of the core business process of an educational institution and defines the IS that contributes to the optimizing of the business process. EA essentially focuses on strategy and integration. MDA makes use of models as its main element while focusing on efficiency and quality. SOA on the other hand, depends on service as its key element while concentrating on flexibility and reuse.

The purpose of this study is to formulate an information technology architecture model that provides clearly defined directions with regard to inputs and outputs for EA development activities. EA captures a HEI as-is and to-be capabilities using a number of models and required to develop the models further. The MDA drives the models that provided from the previous activities. This proposed model specifically emphasizes on IS development by applying MDASOA to ensure that IS help align the direction of the planning, implementation and control process to remain consistent with the enterprise business strategy. The critical task of this development process are selecting the appropriate model for each phase of EA at the right level of MDA in detail. The information system will therefore meets the quality assurance (QA) standard in order to improve the educational quality of HEI in Indonesia [4]. This model will then be tested through a case study by applying the model for University of Lampung (Unila), a public university in Indonesia.

The rest of the paper is organized as follows: section 2 presents the related work. Section 3 gives key technologies and concepts. Section 4 presents analysis of WIS development in educational institution. Architectures used for creating the foundation for WIS development are described in section 5. Section 6 presents WIS development as a case study in Unila including WIS implementation and evaluation, while concluding remarks and future work are provided in section 7.

15.2 RELATED WORK

Integration between business and IT is a major challenge that industries, including higher education institution must deal with. The adoption of the EA concept in this case becomes an absolute need [5]. Various models in different abstraction levels required by EA can be supported by MDA

[6]. The author in [7] has described that the MDA approach can support modeling at hierarchical systems employed for aligning business and IT within an organization. The research also featured techniques applied to design hierarchical systems and developed an integrated enterprise model. A number of studies have adopted the MDA and EA frameworks to tackle modeling issues, such as [8] that described the conceptual mapping between the MDA and Zachman framework.

For web information system (WIS) development, several studies have adopted MDA such in [9] that presented the MIDAS framework. These studies focused on the structural dimension of WIS. Still adopting this framework, authors [10] have complemented the issue on navigation model construction from a user service-oriented perspective. In addition to the functional and navigational requirements of WIS, research that discussed on architectural features resulting in web specifications includes [11]. The authors have described WebSA (Web Software Architecture) based on the standard MDA. WebSA provides a set of architectural models and transformation models to specify a web application. A combination of the MDA and SOA approaches for IS is adopted in [12]. The authors have described Service-Oriented Development Method (SOD-M) that consists of models and the rules for mapping from the business view to the information view. SOD-M distinguishes between business modeling and IS in which the IS will be developed. Business view focuses on the requirement of the business, while IS focuses on the functionalities and processes.

Nevertheless, the simultaneous adoption of the EA, MDA and SOA concepts for producing an information technology architecture model for HEI is remain few. This study shall use the approach to generate an information technology architecture model and also specifically emphasize on WIS development especially for HEI to ensure that WIS has a coherent planning, implementation and control mechanism in place consistent with the business strategy of the educational institution.

15.3 KEY TECHNOLOGIES AND CONCEPTS

15.3.1 WEB INFORMATION SYSTEM DEVELOPMENT

The rapid pace of IT development has led to an increasing number of information systems being transformed into web information systems (WIS) [13]. WIS is an effective platform for collecting, storing, managing and disseminating information similar to the function of traditional information systems. The difference however is that WIS can handle vast amounts of information from diverse sources undergoing fast-paced technological advancements, in different formats with high levels of complexity. This web-based system can be integrated into other WIS or non- WIS for enterprise purposes such as integration with database. Recently WIS has become a complex enterprise application. WIS normally integrates existing systems by using interactive interfaces, handles a large number of users with different access rights, and accessible through various of devices. From its development aspect, web-based systems are now more than simply about visual and user interface design. A systems developer must be familiar with the environment and current needs, and armed with the latest approaches that allow users to adapt quickly to the system. As such, WIS development is a complex task that requires approaches and effective tools to assist the developer.

15.3.2 IS : THE CASE OF HIGHER EDUCATION INSTITUTION INDONESIA

Each HEI has its own set of strategies built on the stated vision and mission. IS must therefore be well planned to complement this strategic direction.

IS that is intended in this paper refers to webbased information system or non web-based. Pursuant to government regulations in Indonesia the IS established by university must be prepared to support academic program management and quality improvement. The IS must help HEI meet its QA standard, that should at least consists of data collection, analysis, storage and retrieval, data and information presentation, and communication with relevant parties [4]. The IS evaluation standard in HEI is regulated by the government and covers several key elements, including the availability of the following components [14]:

1. Blue print on IS development, management and utilization, including systems that regulate data flow, data access authorization and disaster recovery systems.
2. Decision support systems to support top management for better planning and self-assessment analysis and more objective decision-making.
3. Database that at least should consists of information on finances, assets, facilities and infrastructure, academic administration, student and alumni profile, and teaching staff and supporting personnel.
4. IS intended for campus internal and external communication, and access to sources of academic information for students and teaching staff.
5. Internet capacity with adequate bandwidth ratio per student.

From the aforementioned elements, information systems that supports HEI activities at a minimum should encompass academic, human resource, financial and asset information systems that ideally web-based platform. An integrated WIS is crucial for the academic community to ensure easy access to data and information for learning, administrative and reporting purposes. Management also requires an integrated WIS to support the decision-making process and for monitoring and evaluating HEI performance. Given the purpose of a HEI, and its WIS requirements along with the business process involved, it can be concluded that the data and information characteristics of a HEI are as follows: (1) distributed in every working unit and typically have different structures and standards

(2) accessed by all members of the academic community with different needs, roles and level of knowledge (3) data is obtained from historical and operational data of the higher education institution (4) continued rapid data growth (5) immense data volume as academic data must be stored for a lengthy period (6) data must be periodically transferred to a different system as it has become input for another system (7) data transactions with varying time periods depending on academic calendar, peak time for certain periods such as the new academic year (8) for data reporting purposes to the central government, academic data must be based on data structure and relations set forth in the National Higher Education Database (PDPT) Data Glossary (determined by the Directorate General for Higher Education, DGHE) [15].

15.3.3 CONCEPTUALIZATION OF INFORMATION TECHNOLOGY ARCHITECTURE MODEL

Compatibility between IS application and institutional needs can be assured by taking into account the integration factor when developing the system. The essential purpose of integration is to narrow gaps found in the systems development process. Enterprise architecture helps reduce these gaps by providing model-based IS planning, designing and managing within the enterprise. With regard to HEI, the EA concept is highly relevant to be applied for maximizing the benefits of having an IS in the educational institution. The IS can therefore strategically increase HEI comparative advantage. EA in general has the following domains: Business Architecture, Information System Architecture, Technology Architecture. Business Architecture defines the business strategy, management, organization and business process. Information System Architecture comprises Data Architecture that describes data structure, data management and resources; and Application Architecture that defines applications required for managing data and supporting business functions. Technology Architecture represents software application infrastructure to support application development. EA has various frameworks and those that are often applied include Zachman Enterprise Architecture Framework [16] and The Open Group Architecture Framework (TOGAF) [17]. These frameworks essentially serve the same purpose of facilitating the design of the IT architecture of an enterprise. Nevertheless, preference over a particular framework also needs to be consistent with enterprise needs [18].

A key factor of EA is the compatibility between business and technology available for all stakeholders. This can be attained when all involved parties share a common perception. In this regard, modeling becomes a crucial aspect that needs to be taken into account [19]. Several models being applied however are still difficult to understand and have backgrounds that are not commonly used. Under such circumstances, the advantages of the preferred framework may not be optimally realized. A model refers to systems specification and normally presented through illustrations and texts. The ability to develop and transform models at different levels is a critical feature of MDA. The MDA is developed by The Object Management Group (OMG) [3]. The main phases of MDA process are analysis, design and model-driven implementation. This corresponds to the Computation Independent Models (CIM), the Platform Independent Models (PIM) and the Platform Specific Models (PSM). The TOGAF Architecture Development Method [ADM] [20] and MDA are mutually complementary and present immense business potential if effectively applied in combination. Hence, synergies between TOGAF and MDA will lead to improvements for the organization particularly for developing better architecture quality. From 9 phases of TOGAF ADM and requirement managements phase, the 6 specific phases have applicable and useful criteria to map to MDA. The applicable criteria are phase A (Architecture Vision) to CIM and PIM levels of MDA, phase B (Business Architecture) to CIM and PIM levels, phase C (Information System Architectures) to PIM level, phase D (Technology Architecture) to PSM level. The useful criteria are phase E (Opportunities and Solutions) and phase G (Implementation Governance) to PSM level. The proposed model in this study based on this approaches [21].

The solution to apply MDA in sync with other frameworks in order to establish the required model is expected to address any existing modeling issues. The model-driven method is a systems development approach in which the model is described in a way that clearly defines system comprehension, design, development, application, maintenance and modification. MDA in this case is intended to develop an information system that can be applied to describe enterprise business and resources. Through this form, the enterprise will gain the ability to generate specific applications and if required, make the necessary modifications according to changing needs which will later be represented in code form. Similarly, using SOA will complement the WIS being developed. Web services, as the applied technology, on the other hand shall regulate on how the system will interact and be utilized by WIS or other applications. To effectively implement SOA, it is therefore crucial to ensure an accurate analysis of the required data, ongoing business process, as well as applications and interfaces employed in running the business process in every unit within HEI. The simultaneous use of all of these approaches shall lead to a welldefined information technology architecture understandable to all stakeholders, while MDA-SOA helps ensure a more effective and manageable systems development process.

15.4 ANALYSIS OF WIS DEVELOPMENT IN EDUCATIONAL INSTITUTION

In the academic world, information serves as one of the most valuable resources that need to be well managed in order to accomplish goals set by the educational institution. Education-related information as vital resource for the educational institution includes content and curriculum, learning process, facilities and infrastructure, and human resource. This indispensable resource should ideally complement and support the existing business process. In reality however, HEI in Indonesia needs to anticipate and deal with several of the following persistent issues in information systems such as redundancy, lack of standardization, lack of consolidation and inconsistency. Therefore the development strategies, requirements, issues and challenges needs to be analysed to determine the needs of educational institution.

15.4.1 WIS DEVELOPMENT STRATEGY

Given the importance of WIS development, HEI as an implementing enterprise must draw up clearly defined guidelines and plans for developing a comprehensive WIS within its organization. Based on an analysis of the overall situation of the institution along with its general information system characteristics, the approach required to establish WIS within HEI must have the following qualities (1) able to integrate various existing standards and platforms (2) can be easily adapted to changes (3) facilitates the developer in the WIS development process (4) can be easily implemented (5) must have clear guidelines for evaluating WIS implementation.

Many existing information systems have been established in the HEI but are not integrated with the exist of the systems. The strategy to retain existing IS still in operation, re-engineer or replace with new systems can be seen from the system modularity parameter, on whether the existing IS can or cannot be integrated into other IS. If the existing IS is difficult to integrate, this means that it may no longer be viable to maintain the system due to its obsolete technology. Meanwhile, for existing IS that is to be maintained, the appropriate interface needs to be established to facilitate integration with other IS, or interface for the data warehouse. A new WIS is being developed in stages according to the integration standard after which the previous IS can then be terminated.

15.4.2 REQUIREMENTS ANALYSIS OF WIS

Requirements analysis is intended to collect information necessary for WIS development that shall help meet HEI business goals. Requirements are identified based on the internal and external needs of the higher education institution.

15.4.2.1 INTERNAL REQUIREMENTS

Internal requirements are formulated according to the needs of the higher education institution as well as its departments and relevant divisions. The requirement of each department and division may differ but in general will be adjusted by conforming to the existing procedure. Given the internal requirements, the following aspects should be taken into account during IS development:

- Web-based platform
- Supports changes including with regard to changes in the curriculum, courses, codification or transcript
- Allows automation of academic transactions
- Allows the customization of functions specific to the needs of departments and working units

15.4.2.2 EXTERNAL REQUIREMENTS

External requirements are formulated based on government regulations applicable in Indonesia which in this case refers to the Ministry of Education, higher education policies and strategies, reporting requirements and other general requirements. Given these external requirements, the WIS to be developed must:

- Be compatible with the National Higher Education Database (PDPT)
- Support standards and formats for academic program reporting purposes to the government (DGHE) such as the EPSPED report (study program evaluation based on self-assessment) [22]
- Comply with QA standards for the accreditation of HEI issued by the National Accreditation Board for Higher Education (BAN-PT).

15.4.3. ISSUES AND CHALLENGES

15.4.3.1 INTEGRATION

Higher education institution needs to be encouraged to develop an WIS integrated into a model that is built according the needs of working units within the respective institution or among university. Certain data required by departments and working units as inputs is essentially the same except for differences in its usage and reporting mechanism according to their respective business process. An integrated system shall significantly support interrelated business processes and optimize its use. For example, an effective academic system that takes into account the number of students will be capable of predicting the number of required human resource (in relation to human resource system), incoming and outgoing costs (in relation to finance system), as well as the number of required classes and rooms (in relation to assets system).

To ensure integration with existing systems, new technologies should be combined with existing technologies with the least possible effort without having to redevelop ongoing systems. Based on the current situation of the existing IS in university, integration may come in the following forms:

- Data integration of different business processes, different IS functions and processes yet related to the overall work cycle, and database integration.
- Availability of an interface that can form linkages among existing IS.
- Users need not login for every WIS in order to access the required data and information. They only need to login once to enable access to all relevant WIS.
- Data that serves as output for a certain process in an IS can be the input for another IS.

15.4.3.2 INTEROPERABILITY

WIS development must consider the interoperability of systems. Interoperability refers to the ability of a system to work in sync with another system that allows information exchange and the ability to use the shared information. With regard to WIS implementation, the technical and regulatory aspect of data exchange needs to taken into account. A key factor related to the technical aspect is high-level interoperability to ensure that data transfers from source to target can be done regardless of the diversity in hardware and software platforms. An open architecture solution is therefore necessary to allow the smooth exchange of data and information from a different system that works in sync with systems inside and outside of university. Apart from technical issues, interoperability also needs to be regulated through the appropriate policies. This is essential in order to ensure uniformity in format and data exchange mechanisms among HEI for the purpose of guaranteeing high-level interoperability.

15.5 ARCHITECTURE USED: CREATING THE FOUNDATION FOR WIS DEVELOPMENT

Based on the previous analysis, this section shall explain on the information technology architecture model which serves as a reference model for WIS planning, design and organizing within HEI. The proposed EA-MDA model exhibits all activities required to develop the architecture for HEI, beginning from vision construction to implementation and evaluation. The WIS development process will specifically be described in detail by adopting the MDA-SOA approach. Figure 1 presents the proposed model.

Formulating the information technology architecture model for HEI is inextricably linked to the analysis of the university internal and external business functions as regulated by the government and set out in regulations on higher education in Indonesia. Business function is divided into two groups: primary business function and supporting business function. The primary business function of universities in Indonesia generally comprise primary business activities grounded in the Triple Principles of Higher Education Institutions, namely education, research and community service. The supporting business function on the other hand contains supporting business activities that include academic administrative management, human resource management, finance management and asset management. All of these activities must meet the QA and Executive Information standards.

15.5.1 ENTERPRISE ARCHITECTURE

15.5.1.1 BUSINESS ARCHITECTURE

Business architecture in general describes the series of business activities, data and information found in the internal and external environments of

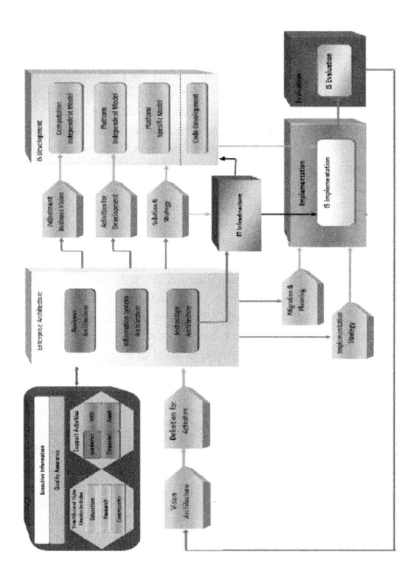

FIGURE 1: Information Technology Architecture Model

the HEI. This phase is specifically meant to gain insight on the current situation of the business process before improvements are recommended by initiating the business architecture modeling process. HEI business architecture modeling may refer to HEI business functions already analyzed previously specifically by considering the primary and supporting business functions of the higher education institution. Business functions can be identified by using Porter's value chain approach. At this stage, modeling tools and basic methods such as Business Process Modelling Notation (BPMN) [23] and Unified Modeling Language (UML) [24] can be employed to develop the desired model.

15.5.1.2 INFORMATION SYSTEM ARCHITECTURE

15.5.1.2.1 DATA ARCHITECTURE

Data architecture refers to the identification of data that supports business functions as defined in the business model explained earlier. Management requires centralized and integrated data sources in order to heighten coordination and synchronization of data management operations. Within the context of data architecture, university needs to avoid the repeated inputting of the same data. Data should be inputted only once and can subsequently be used simultaneously by relevant business processes. The data architecture currently developed by university still has weaknesses that need to be addressed before it can be integrated with data originating from other universities due to lack of integration in their data structures.

In this stage, data architecture is developed through the identification of the business function and organizational entity. Results of the identification process are then presented through an UML class diagram or a data functional matrix to illustrate the connection between business process and data entity through created, use, read and delete (CURD) functions. The relation between each data entity with another entity is analyzed then compared with the list of data currently being managed by the system. This comparison is necessary to ascertain on whether data covered in the data architecture entity is based on data description.

15.5.1.2.2 APPLICATION ARCHITECTURE

Application architecture focuses more on the planning of application needs and creating of application models. Application architecture required by the HEI is an integrated, online application that runs on a standard platform. In addition, management also requires a dynamic application and real-time system to ensure the presentation of timely and accurate information. Application architecture can be described through the application interaction matrix of business functions within the organization or the technical reference model (TRM). Application is linked to business and organizational functions in order to keep track of the collective use of an application. The impact of application architecture towards existing applications is analyzed. The appropriate solutions are required in order to determine on whether available applications are to be maintained or modified, integrated into other applications or new applications developed.

15.5.1.3 TECHNOLOGY ARCHITECTURE

Technology architecture seeks to identify technology platforms, analyze the use of current technology platform toward applications, and propose technology platforms related to university needs. The appropriate technology platform is selected by assessing current IT trends and developments such as trends in hardware, software, network, database, security and social network. Results of technology classification include the selection of viable technologies for technology platforms that shall support applications and the recommended technology development. The technology architecture model essentially describes how technology supports applications and user interaction when using the applications.

15.5.2 SUPPORT ACTIVITIES

15.5.2.1 VISION ARCHITECTURE

This stage involves the identification of management requirements, project scope and constraints, and the definition of advanced architecture and expected targets. These identified aspects are represented through the stated vision and mission, business goals, and business objectives. The output at this stage is the creation of the vision architecture to be used. To have an idea of the general universities vision in Indonesia, several samples of Indonesia's leading universities have been examined. From this sampling it can be conferred that the stated vision of university in general is to establish themselves as leading seats of learning of national and international repute by building on their respective core competencies. Their mission statement on the other hand aspires to deliver first-rate educational processes based on the Triple Principles of Higher Education Institutions.

15.5.2.2 DEFINITION FOR ACTIVITIES

Prior to starting work on the subsequent architecture model, it is necessary to document and define all data needs, data organizing applications and data sharing requirements in running the business process, and existing technology platforms employed within the enterprise. Documentation shall become the basis for architecture modeling and the following implementation plan. Based on the primary and supporting activities of the higher education institution, this stage shall define activities necessary for documentation purposes that cover literatures, surveys and interviews with relevant parties.

15.5.2.3 ADJUSTMENT BUSINESS VISION

Architectural models obtained in the previous stages will be reviewed at the next activity. The modeled business and application architecture in the previous stages is then re-assessed with regard to its compatibility with university business vision. This stage maps out the relation between application architecture and the achievement of the vision formulated earlier. The output of this activity will be the basis for the development of the system, especially at the CIM level in the MDA approach.

15.5.2.4 ACTIVITIES FOR DEVELOPMENT

WIS development is also conducted based on the guidelines from the list of WIS development activities. This list is necessary to provide a more focused WIS development process, beginning from preparations for needs assessment to the eventual development of a systems design for use by the higher education institution. The output of this activity will be the basis for the development of the system, especially at the PIM level in the MDA approach.

15.5.2.5 SOLUTION AND STRATEGY DEVELOPMENT

This activity is intended to identify main issues and seek viable solutions. From the results of the analysis on the current situation of the higher education institution and the architecture model produced, a gap analysis is conducted. Results of this gap analysis can become the solution and strategy for resolving issues. This stage is also performed to determine the priority of application development. These solutions and strategies in turn can serve as input for the subsequent process of developing a more comprehensive WIS and technology infrastructure need.

15.5.2.6 MIGRATION AND PLANNING

The migration strategy for a new WIS is required prior to the implementation. In this stage, a gap analysis is essential on the resource base, including changes that may arise upon the implementation of the system. In addition, it is also equally important to conduct an impact analysis of the new application architecture and the decision-making process towards new IT investments in order to ensure a complete planning process with regard to systems migration and implementation.

15.5.2.7 IMPLEMENTATION STRATEGY

This stage involves estimations on human resource, implementation schedule, costs and benefits of the plan and time required. This includes the formulation of recommendations for each development implementation, determining the implementing organization and guaranteeing the compatibility of systems development with the desired architecture. Risk management is another essential component that needs to be incorporated into the implementation plan in order to minimize potential risks, such as risks related to human resource, unmet implementation schedule and others.

15.5.3 TECHNOLOGY INFRASTRUCTURE

The gap that might occur between the current architecture and propose architecture can be eliminated by means of fulfilling the needs of technology and information system. This stage is conducted through procurement for technology support needed to encounter the needs of the technology of addition or optimization technology.

15.5.4 MDA-SOA BASED WIS DEVELOPMENT

The web information system development process is based on the MDA-SOA concept. Metamodeling defines the structure of models, and in an abstract way specifies the construct of a modeling language and their relations. Through this approach, it is necessary to determine a metamodel that suitably represents the web application. During the design phase, modeling is used to define requirements and provide model details at various levels.

Next, model transformation is carried out in accordance with transformation rules. The models provide support for testing prior to implementation and also contribute to automatic code generation.

15.5.4.1 COMPUTATION INDEPENDENT MODEL LEVEL

This stage is associated with an enterprise architecture particularly business architecture that already proposed in the previous stage. Business models need to describe the environment in which the system will be used, with no direct orientation on how it will be implemented. It also specify the requirements, use cases and the system's main flow from the customer's perspective.

15.5.4.2 PLATFORM INDEPENDENT MODEL LEVEL

Modeling the systems process and structure can be done through these models in an independent way of the technological details for their implementation. The processes are captured by UML activity diagram and structures are capture by UML class diagram. These models can be modified or transformed into PSM that express the specific implementation process. For this particular study, the Model View Controller (MVC) pattern is applied as a target technology for the generated web application. Templates need to be defined in order to automatically generate the web application and conduct model transformations. They state transformation rules for changing a given model-to-model or model-to-text transformation. These templates are designed to implement the classes for Data Definition Language (DDL) scripts, Web Service Description Language (WSDL) and for the MVC_Controller of the MVC pattern that functions to receive requests, invoke the MVC_Model to perform the requested operations and send data to the MVC_View. MVC_View formats are to be presented in a web application as PHP files output. The defined templates also facilitate the generation of highly specific PSM from PIM.

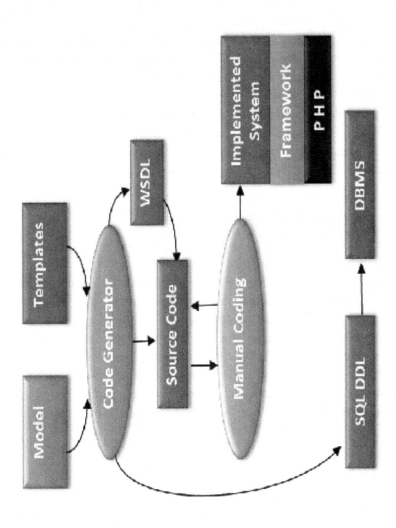

FIGURE 2: Detail of WIS development process

15.5.4.3 PLATFORM SPESIFIC MODEL LEVEL

Based on the models and templates identified earlier, the code generation process can proceed to automatically produce the source code. In this case, Acceleo [25] will be applied for creating the templates and producing a PHP-based web application [26]. Web application is implemented through the PHP scripting language and additional technologies, such as the Apache web server and MySQL database server, as the target environment for deployment, while the adoption of the CodeIgniter [27] PHP framework is based on the MVC pattern. All files generated are placed into the framework. The remaining part are created manually to finalize the entire process. The detail of WIS development steps as shown in Figure 2.

15.5.5 IMPLEMENTATION

The implementation stage is where the WIS already developed in the previous phase is implemented. Through systems implementation, the outcomes can be operated and utilized optimally according to needs. At this stage, it can be determined on whether the system that has been developed can operate according to plans, and whether it can be utilized by the end user and meet the expected objectives. An evaluation is also conducted during this stage on shortcomings during the system development process, including a deficient requirements analysis, not modeled and unpredictable factors. As such, the necessary improvements can then be made to rectify these inadequacies.

15.5.6 EVALUATION

Evaluation is an important aspect necessary for gleaning information on the extent to which the objectives of an WIS has been successfully achieved. Through evaluation, feedback will also be generated, crucial for improving WIS quality in the future. The main constraint in conducting evaluation is in determining the evaluation criteria, evaluation parameter and methodology for establishing the evaluation framework. Recent developments in information technology, particularly with regard to the internet, have made it even more difficult to gauge the level of success and effectiveness of an WIS compared to earlier conditions. WIS has brought forth a different set of conditions and factors that need to be considered, such as direct user interaction unrestricted by time, distance and place.

From previous studies, several recommended models for IS evaluation include Technology Acceptance Model (TAM) and IS-Impact. TAM introduces two key variables - Perceived Ease of Use (PEU) and Perceived Usefulness (PU)—that have central relevance to predict user acceptance [28]. Through the IS-Impact model, IS impact can be measured in terms of Information Quality (IQ), System Quality (SQ), Individual Impact (II) and Organizational Impact (OI) [29]. This however does not provide a comprehensive evaluation of Web-based information system. In view of this, the evaluation model for IS implementation is modified from the aforementioned models by incorporating specific measurements. The selection of the two existing models and the necessary modifications are essential with regard to WIS for the higher education institution in which the characteristics are different from other enterprises.

The defined dimensions for measuring WIS by adding 4 dimensions: User Capabilities (UC), Organisation Capabilities (OC), Information Effectiveness (IE) and System Effectiveness (SE). Based on the characteristics of the HEI, the dimensions used still needs to be adapted for specific user. For example, the OI dimension is not provided to the user student due to not directly related to the organization impact. The elements for measuring the dimensions consist of 19 elements for students, teachers and related working units user, 10 elements for System Developer, 10 elements for Managements as shown in Table 1.

15.6 WIS DEVELOPMENT: A CASE STUDY

This section shall discuss on the application of the information technology architecture model recommended above to support WIS development at Unila as the case study. Unila is a stateowned university in Indonesia that runs 8 faculties, 54 undergraduate study programs (diploma and bachelor's degree), 12 post-graduate study programs (master's and PhD degree) and currently serves 28.116 students, 1120 lecturers and 707 administrative staff (data from academic year 2011).

TABLE 1: Elements for measuring the dimensions

User	Element
Students, Teachers, Staff, related working units	Enjoyment, Number of site visits, Number of transactions executed, Availability, Reliability, Accessibility, Response time, Ease of use, Ease of learning, Navigation patterns, Functions, Training, Services, Empathy, Responsiveness, Assurance, Security, Integration, System features
System Developer	Conciseness, Timelines, Consistency, Maintainability, Applicability, Server and Network Speed, Traceability, Security, Currency, Interactivity
Management	Cost reduction, Time savings, Improving work efficiency, Enhancement of communication, Enhancement of coordination, Improved decision making, Completeness, Relevance, Number of academic standards procedure supported, Number of operational university supported

Data and analysis on university activities related to its Business Function, Vision Architecture, and Definition for Activities required as input for modeling Unila's enterprise architecture are entirely based on current data and conditions necessary for Unila. Due to limited space, not all will be presented in this paper.

15.6.1 BUSINESS ARCHITECTURE

For the modeling of the business process, Unila's business functions are identified and its organizational structure documented. These identified business functions are then linked to the working units in order to determine the responsibilities of working units in relation to a business function. Business functions are identified by using the value chain model that classifies business areas into primary activities and supporting activities of the enterprise. From this model, it can be concluded that Unila has five supporting activities and three primary activities.

Three primary activities related to education, research and community service while five supporting activities related to human resource, finance, asset, services and public relation management.

15.6.2 INFORMATION SYSTEM ARCHITECTURE

15.6.2.1 DATA ARCHITECTURE

Data architecture consists of both existing and planned data architecture. Data architecture draws from the business architecture explained earlier which refers to data required for facilitating three primary activities and five supporting activities.

15.6.2.2 APPLICATION ARCHITECTURE

Application architecture refers to current applications (as-is) and planned applications (to-be). An analysis is therefore necessary on ongoing applications to look at the business process and needs of working units. Analysis results on existing applications based on faculties, departments and working units responsible for managing the applications. In the development of current applications, Unila have three developments options, e.g., open-source based development, ownsource based development, and for some special purpose the application already be developed by the government and ready to be implemented. Available application technologies mostly use PHP and database and Oracle. A few other technologies still operate through Novel Netware by using DBF database. Websites use PHP and MySQL.

Analysis results on existing applications inform decisions on strategies to be implemented for WIS integration with the institution. Subsequently, to identify future application needs, a Strengths, Weaknesses, Opportunities, and Threats (SWOT) and Critical Success Factors (CSF) analyses are conducted to determine the future WIS solution model appropriate for the business process carried out by Unila. Figure 3 shows the analysis results of planned application (to-be) required by Unila. There are classified in the Academic management, Knowledge management, Resource management, IT Services, and Community Relationship Management (CRM) to support Quality Assurance and Executive information.

15.6.3 TECHNOLOGY ARCHITECTURE

This particular phase is intended to identify the current technology platforms and how they are applied with regard to applications, and determine the recommended technology platforms that Unila requires. Unila's existing technology platforms identified include its data center equipped with 32 servers, 2000 desktop computers connected to the Local Area Network (LAN), and data storage in the form of SAN (Storage Area Network). Its intranet infrastructure that links internodal working units relies on a switching system applicable up to the faculty level whereby all switching functions are done through the manageable switch mode. Fiber optic as the main media backbone has 1 Gbps in channel bandwidth. Wireless media is also used as the network infrastructure, although several study programs (30% from all study programs in Unila) are still using cables and have yet to employ wireless technology. The existing technology platforms are largely support of the proposed application but still need optimization to improve the performance. From Unila's needs assessment, the technology required by the institution is a network technology that connects applications and allows easy access for end users. Technology platforms recommended for development must be those that adopt recent technological trends such as the latest in hardware, in memory computing (IMC), network, cloud computing, web 2.0., e-Learning 2.0, mobile computing, Internet of Things (IOT) and social network.

15.6.4 SOLUTION STRATEGY FOR BRIDGING THE GAP

Subsequent activities as follow up to architecture modeling include Adjustment Business Vision, Activities for Development, and Solution and

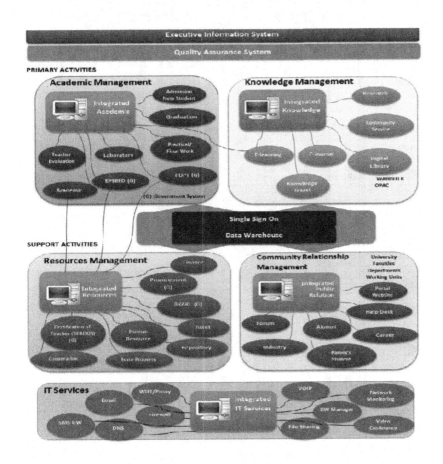

FIGURE 3: Application Architecture (to-be)

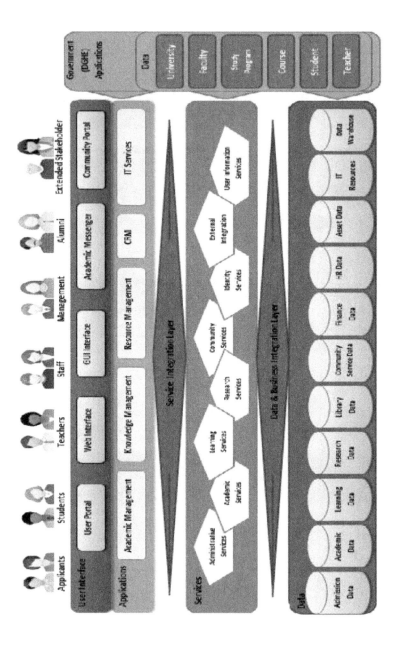

FIGURE 4: Integration Model Architecture

Strategy Development. Based on the current situation of existing IS in Unila and HEI in general, solutions and strategies are formulated for integration purposes with regard to the presentation (user interface), applications, services and data. The integration is also perform with the data on DGHE (government application). Figure 4 presents the integration model architecture.

This particular phase also determines the list of priority for application development. The solution model produced can connect existing applications with other ongoing applications as well as with new applications without affecting the performance of the respective application. Concerning WIS development, based on Unila's needs assessment, the Academic management domain shall be given the highest development priority. The following example focuses on the development of a web information system (called SIMPEL) [30] to integrate the existing systems provided in the analysis results mentioned in section 6.2.2.

TABLE 2: The templates

Template Name	Description	Total Template
MVC pattern	A template for generating Model, View (user interface) and Controller	10
DDL	A template for generating DDL scripts to create the tables	1
WSDL	A template for generating WSDL files	1

15.6.5 APPLYING MDA-SOA BASED WIS DEVELOPMENT

In this work, a set of models is designed related to the WIS. A specific metamodel acquaintance of UML metamodel is created for this modeling process. Based on the result obtained from the modeling process, this process expects to generate artifacts that may help create the necessary web pages. Nevertheless, the modeling process only constitutes half of the entire MDA system. Templates to produce the target system are equally important. The usefulness of the templates is to transform a given model to text as shown in Table 2.

By using the defined templates and models, the generator code creates all application source code based on MVC pattern, DDL and WSDL files.

The SIMPEL development in this work realized 107 FPs (Function Point). Handwritten (manual) codes need to be added after generation for complex business logic and graphical user interface (GUI) parts. Then the result of generated and handwritten code are tailored to the framework for system implementation. The development of WIS is combined with a framework CodeIgniter, leading to faster PHP web application development.

15.6.6 WIS IMPLEMENTATION

From the results obtained in the previous stages, the proposed WIS has not been entirely developed and implemented. Similarly, the proposed technology platform. This case study is taken to integrate two existing WIS by using a WIS that serves as an interface. The development process involved in implementing the approach allows developers to progress smoothly from requirements specification to the generation of the web application. The implementation of a WIS has successfully integrated Course Management System (CMS) MOODLE [31] into Unila's academic information system and work properly. The implemented SIMPEL is analyzed in the teaching/learning activities at Department of Electrical Engineering Unila on trial period in odd and even semester of the 2011/2012 academic year. During this period a total of 887 students were enrolled in 25 courses through the SIMPEL and follow the learning activities in MOODLE.

15.6.7 WIS EVALUATION

The model adopted for WIS evaluation is a modification of the IS-Impact and TAM models with 12 dimensions of measurements. This model is applied to assess the quality and impact of SIMPEL implementation in Unila. The designed questionnaire is consists of 19 elements for user students and teachers. Given the time constraint in evaluating the implementation of SIMPEL, the evaluation process only involves participants representing user students and teachers and has yet to involve all stakeholders in Unila.

Some 142 users have expressed their views through the questionnaire on SIMPEL implementation for the recommended dimensions. Results as shown in Table 3 reveal that 69.2% of users are satisfied with the SIMPEL (highly agree and agree), neutral 24.6% and 6.2% unsatisfied (disagree and highly disagree).

TABLE 3: Evaluation result

Element	Highly agree	Agree	Neutral	Disagree	Highly disagree
Enjoyment	64.9%	26.6%	7.1%	1.4%	0.0%
Number of site visits	62.1%	30.5%	5.7%	1.8%	0.0%
Number of transactions executed	32.1%	45.6%	19.2%	3.2%	0.0%
Availability	20.3%	50.4%	25.6%	3.4%	0.4%
Reliability	7.8%	48.4%	27.4%	3.5%	12.9%
Accessibility	5.7%	43.8%	32.7%	5.3%	12.5%
Response time	6.4%	33.4%	54.5%	5.7%	0.0%
Ease of use	7.4%	62.7%	26.0%	3.9%	0.0%
Ease of learning	20.7%	52.0%	25.3%	2.1%	0.0%
Navigation patterns	7.1%	64.8%	25.3%	2.8%	0.0%
Functions	20.0%	53.8%	19.1%	6.9%	0.1%
Training	18.2%	49.8%	15.1%	4.4%	12.5%
Services	17.5%	30.4%	36.1%	3.5%	12.5%
Empathy	7.1%	60.2%	29.5%	3.2%	0.0%
Responsiveness	18.5%	31.3%	45.2%	4.6%	0.4%
Assurance	30.1%	40.6%	21.9%	7.3%	0.2%
Security	22.1%	57.8%	19.0%	1.1%	0.0%
Integration	26.2%	59.8%	13.1%	0.9%	0.0%
System features	7.9%	71.8%	19.8%	0.5%	0.0%

Questionnaire results also show that the significant elements in SIMPEL are Number of site visits, Enjoyment and Integration. It show that SIMPEL effectively assist the teaching learning process. While the elements that must be improved are Accessibility, Training and Reliability. It shows that SIMPEL still need to be complemented and enhanced the quality.

15.7 CONCLUSIONS

Based on an analysis of IT/IS of HEI in Indonesia, an EA-MDA-based information technology architecture model has been developed according to the situation and needs of national universities. This model is then applied to University of Lampung as a case study. The outcome is an architecture information technology that can be implemented to support integrated WIS development in Unila. One of the WIS developed in Unila has been implemented and subsequently evaluated which resulted in an impressive level of user satisfaction. This study is expected to specifically contribute to the advancement of HEI in Indonesia, particularly as reference for designing and building a comprehensive WIS. The definition of a complex architecture model shall help meet all university needs. Nevertheless, HEI need not force themselves to create unnecessary business processes, but it would be best for them to focus on areas most suited with their respective needs. The adoption of the MDA therefore substantially supports this concept. The future work we will find other strategies in addressing issues of the integrated WIS and focus on improve the models in each activities prior to developed.

REFERENCES

1. Osvalds, Gundars, (2001) Definition of Enterprise Architecture – Centric Models for The Systems Engineers, TASC Inc.
2. T. Erl, (2005) Service-Oriented Architecture - Concepts, Technology, and Design, Prentice Hall.
3. OMG, "OMG Model Driven Architecture," Available: http://www.omg.org/mda.
4. Republic of Indonesia, (2005) "Government regulations (PP) No.19 of 2005 of national standard of education," Jakarta, Indonesia.
5. Shah, H. & Kourdi, M.E (2007) "Frameworks for Enterprise Architecture," IT Professional , vol.9, no.5, pp. 36-41
6. Ostadzadeh, S. S, Aliee, Fereidoon S., Ostadzadeh, S. A., (2008), "An MDA-Based Generic Framework to Address Various Aspects of Enterprise Architecture", Advances in Computer and Information Sciences and Engineering, pp. 455–460.
7. Wegmann, A. and Preiss, O., (2003) "MDA in Enterprise Architecture? The Living System Theory to the Rescue", Proceedings of the 7th IEEE international Enterprise Distributed Object Computing Conference (EDOC'03), pp. 2-13

8. Fatolahi, Ali, S. S, Stéphane, C. L, Timothy (2007) " Enterprise Architecture Using the Zachman Framework: A Model Driven Approach", Proceedings of the IRMA 2007 International Conference, pp 65-69.
9. Cáceres, P., Marcos,E.,Vela ,B., (2003), "A MDA-Based Approach for Web Information System Development", Workshop in Software Model Engineering (WISME'03)
10. Castro,V., Marcos, E., Cáceres, P., (2004) "A User Service Oriented Method to Model Web Information Systems", WISE 2004, LNCS 3306, pp. 41–52
11. Meliá, S., & Gómez,J.,(2005)" Applying Transformations to Model Driven Development of Web Applications ",ER Workshops 2005, LNCS 3770, pp. 63-73
12. Castro, V. De., Marcos, E., and Vara, J.M., (2010), "Applying CIM-to-PIM model transformation for the service-oriented development of information systems," Journal Information and Software Technology, pp. 87-105
13. Carstensen, P. H., Vogelsang, L, (2001), "Design of Web-based information systems—new challenges for systems development?" Ninth European conference on information systems (ECIS), pp 536–547
14. National Accreditation Board for Higher Education (Badan Akreditasi Nasional Perguruan Tinggi, BAN-PT) Ministry of Education Indonesia, (2008) Book IIIA: Undergraduate program study accreditation, Jakarta, Indonesia, BAN PT.
15. Directorate General of Higher Education (DGHE), "Dissemination of national higher education database system", Ministry of Education Indonesia, (2010), Jakarta, Indonesia.
16. Sowa, J. F., Zachman, J.A. (1992) "Extending and formalizing the framework for information systems architecture". IBM Systems Journal 31, No. 3, 590-616.
17. The Open Group, TOGAF Version 9.1, http://www.opengroup.org/togaf/
18. Leist,S., & Zellner,G.,(2006), "Evaluation of current architecture frameworks". In Proceedings of the 2006 ACM symposium on Applied computing
19. Frankel, David S.,(2003) The Model Driven Architecture: Applying MDA to Enterprise Computing, OMG Press.
20. Open Group, (2009) The Open Group Architecture Framework: Architecture Development Method.
21. Rivett,P., Spencer,J., Waskiewicz,F., (2005) TOGAF/MDA Mapping, White Paper, The Open Group
22. Directorate General of Higher Education (DGHE), (2009), EPSBED, http://evaluasi. or.id/
23. The OMG Group, (2008) Business Process Modelling Notation, Version 1.1.
24. Booch, G., Rumbaugh, J., Jacobson, I., (1999) "The Unified Modeling Language: A User Guide", Addison Wesley.
25. Main Acceleo Site, "Acceleo MDA Generator," http://www.acceleo.org/.
26. Main PHP Site "PHP," http://www.php.net/.
27. Main CodeIgniter Site , "CodeIgniter," http://codeigniter.com/.
28. Chuttur, M.Y., (2009), "Overview of the Technology Acceptance Model: Origins, Developments and Future Directions", Indiana University, USA, Sprouts: Working Papers on Information Systems
29. Gable, G.,Sedera, D.,Chan, T. (2008), "Re-conceptualizing Information System Success: The ISImpact Measurement Model", Journal of the Association for Information Systems, 9 (7), pp. 377-408.

30. Mardiana & Araki,K., (2012) "SIMPEL: An innovative web application interface supporting online course management system," in Information Technology Based Higher Education and Training (ITHET), pp.212-219.
31. Main Moodle Site, "Moodle," http://moodle.org/

AUTHOR NOTES

CHAPTER 4

Acknowledgments
This material is based upon work financially supported by the National Research Foundation. The researcher would also like to thank Mr Johan van Tonder at Anglo Platinum for his valuable inputs.

CHAPTER 9

Acknowledgments
The authors would like to thank the anonymous referees for constructive comments on earlier version of this paper.

CHAPTER 11

Acknowledgments
This material is based upon work financially supported by the National Research Foundation.

CHAPTER 12

Acknowledgments
The authors would like to thank the Ministry of Higher Education, Malaysia and Universiti Utara Malaysia for providing the research grant for this study. Likewise, we are also indebted to Jaap Schekkerman, President of the Institute for Enterprise Architecture Developments (IFEAD) for giving us the permission to use the IFEAD EA Trend Survey Questionnaire in this study.

CHAPTER 14

Acknowledgments

We would like to thanks the three organizations (anonymity maintained) for their participation in the survey.

CHAPTER 15

Acknowledgments

The first author would like to thank the Directorate General of Higher Education (DGHE) of the Ministry of National Education of Indonesia that has provided the funding of this study. Thanks are also to the colleagues of the Department of Electrical engineering and Computer Centre University of Lampung for their help and support.

INDEX

Printed in the United States
by Baker & Taylor Publisher Services